T0350449

Quantum Reflections

This volume introduces some of the basic philosophical and conceptual questions underlying the formulation of quantum mechanics, one of the most baffling and far-reaching aspects of modern physics. The book consists of articles by leading thinkers in this field, who have been inspired by the profound work of the late John Bell. Some of the deepest issues concerning the nature of physical reality are debated, including the theory of physical measurements, how to test quantum mechanics, and how classical and quantum physics are related. This book will be of interest to students with a background in quantum physics, who wish to explore in more detail its philosophical aspects, practising scientists who are not content with blindly applying the rules of quantum mechanics, and anyone interested in gaining a deeper understanding of the philosophy of physics.

Quantum Reflections

Edited by John Ellis and Daniele Amati

CAMBRIDGE
UNIVERSITY PRESS

University Printing House, Cambridge CB2 8BS, United Kingdom

Cambridge University Press is part of the University of Cambridge.

It furthers the University's mission by disseminating knowledge in the pursuit of education, learning and research at the highest international levels of excellence.

www.cambridge.org
Information on this title: www.cambridge.org/9780521630085

First published 2000

A catalogue record for this publication is available from the British Library

Library of Congress Cataloguing in Publication data
Ellis, John, 1946–
Quantum reflections / John Ellis and Daniele Amati.
p. cm.
Includes index.
ISBN 0 521 63008 8 (hardbound)
1. Quantum theory. 2. Physics–Philosophy. I. Amati, D. II. Title.
QC174.12.E45 2000
530.12–dc21 99-043748

ISBN 978-0-521-63008-5 Hardback
ISBN 978-0-521-08889-3 Paperback

Contents

Contributors

ROGER PENROSE
Mathematical Institute,
University of Oxford

HELMUT RAUCH
Atominstitut der Österreichischen Universitäten

ALAIN ASPECT
Laboratoire Charles Fabry,
Institut d'Optique Théorique et Appliquée

GIANCARLO GHIRARDI
Dipartimento di Fisica Teorica,
Universita di Trieste

JON MAGNE LEINAAS
Physics Department,
University of Oslo

ABNER SHIMONY
Physics Department,
Boston University

KURT GOTTFRIED
Newman Laboratory of Nuclear Studies,
Cornell University

N. DAVID MERMIN
Newman Laboratory of Nuclear Studies,
Cornell University

ROMAN JACKIW
Center for Theoretical Physics, M.I.T.

Foreword

Our wish to honour the memory of John Bell had to reconcile the desire of the scientific community – his colleagues and friends in the first instance – to express their sentiments and admiration, with our wish to respect his attitude of avoiding celebrations and pompous words.

We thus collected from his friends and collaborators this volume of Quantum Reflections in the hope that these papers would have pleased him. John's acute mind and original thinking appear as an underlying theme giving unity and consistency to this collection, in a field where even posing questions has a deep scientific and epistemological connotation.

John Stewart Bell was a unique individual, who pursued with remarkable honesty and consistency a very personal attitude towards life, science and society. Alien to fashion and to easy enthusiasms for trendy ideas, he always remained faithful to his own view of what constitutes 'understanding'. From an early age, he was fascinated by the unsuspected depths of quantum mechanics, into which he saw more deeply than the rest of us could imagine. For these reasons John holds a dear place in our hearts, and in many others.

John would have been very much against a mournful review of his life's work, which is why we chose to assemble in his memory a collection which looks forward to the heights which pygmies might hope to reach by standing on his giant's shoulders. We hope that this collection will be worthy of his memory.

John leaves an impressive contribution to our field of research, together with the memory of his so pleasant personality, which all those who knew him have the privilege to keep. His deep sense of humour accompanied the depth and rigour with which he was approaching all questions in life as well as in science. He impressed us all by the depth of his reasoning. He did not fear to enter intellectual territory where few others dared to tread. He contributed in a magnificent way to a better understanding of quantum theory and brought light and decisive answers to questions which had led even Einstein to scepticism and doubt.

We would like to quote from the book of Viki Weisskopf who, in 'The Joy of Insight', writes about John:

> Probably the most penetrating mind among CERN theorists is John Bell, who is well known for his insistent striving for a better understanding of the foundations of quantum mechanics. In my discussions with him I often had to admit that my ideas about some fundamental principles of quantum mechanics were not sufficiently thought out. He admonished me for my fuzzy thinking with such charm and clarity that I not only learned something but also thoroughly enjoyed the lesson.

Whilst always raising the questions still left open by our present understanding of quantum mechanics, John had much sympathy with those he called the 'why bother?ers', agreeing with them that 'ordinary quantum mechanics is just fine for all practical purposes'. He himself put this approach into practice in particular when, together with Jack Steinberger, he helped cast the definitive formalism for the violation of CP symmetry.

Despite all his honours, his modesty, particularly with respect to the philosophical causes supported by some of his ideas, was impressive. As he himself said: 'What I really wanted was a clean argument rather than justify any particular conception of the world. From what I know of my own character, which is somewhat stubborn, I am often more concerned with the conduct of the debate and its logic than the actual truth.' On his success with his inequalities, he would say 'Then people started doing the experiments. The results confirmed ordinary quantum mechanics and therefore dashed Einstein's hopes. Then there was more and more publicity.'

Despite the publicity, his approach to physics always epitomized depth, modesty, rigour and imagination. In his words again:

> What we deal with in physics are the simplest situations. We simplify questions to the limit in the hope of finding that the laws of simple things can be built up into the laws of complicated things.

We shall long keep him in our minds as an enduring model.

Daniele Amati
John Ellis

Biographical notes on John Bell

(July 28th 1928–October 1st 1990)

The Belfast years

John Stewart Bell was born in Belfast, Northern Ireland. He was the second child of John Bell and (Mary) Annie Brownlee. Both his parents came from large families (9 and 10 children). He had an older sister and two younger brothers. The family had no academic background. Indeed, his father stopped going to school at the age of eight, his parents paying the relevant fines. John was educated first at Ulsterville and Fane Street Public Elementary Schools, and later at the Technical High School, Belfast. There, in addition to the normal school subjects, there were some less academic courses, such as woodwork and bricklaying (theoretical). At the age of sixteen, having completed the school courses and wanting to continue to study, he found that he was too young to enter Queen's University, Belfast. However, he managed to get a job there as a laboratory assistant in the Physics Department, under the supervision of Prof. K. G. Emeléus and Dr. R. H. Sloan. The staff were very helpful there, giving him books to read, and answering questions, so his scientific education did continue. Around this time, influenced by George Bernard Shaw, he became a vegetarian. By now, he was also a 'Protestant Atheist', which he remained all his life. The next year, aged 17, he managed to become a student. In 1948, he graduated with an Honours degree in Experimental Physics, and in 1949 with one in Mathematical Physics. He considered it very good luck to have been a student (one of the last in Belfast) of the refugee Prof. Paul Ewald, the famous crystallographer, and, in general, was extremely satisfied with his education at Queen's University.

Harwell and Birmingham

John left Belfast in 1949 and went to work as a Scientific Officer at the Atomic Energy Research Establishment (AERE) at Harwell, Berkshire, U.K. After a couple of months there he moved to Great Malvern, Worcs., where there was a branch of AERE, to join the theoretical group of W. Walkinshaw. This group was devoted to the design of accelerators, at that time especially linacs. As he had a thorough knowledge of electromagnetic theory, he found the subject rewarding, and contributed, for example, to the mathematical theory of disc-loaded waveguides.

The whole group moved back to Harwell in 1951. The interest then switched to circular machines with the planning for the proton synchrotron for CERN. Again he contributed a number of papers, one of which was an independent derivation of what became known as the Courant–Snyder invariant for the 'strong-focusing' system.

In 1953 he was offered paid leave from AERE to go to study at Birmingham University in the department of the late Prof. Sir Rudolf Peierls. He profited from the supervision there of the late Dr. (later Prof.) P. T. Matthews. While in Birmingham he independently derived the PCT (parity–charge-conjugation–time-reversal) theorem. His proof, which became the first part of his thesis, was on simpler lines than the formal field theory arguments of G. Lüders. The paper of Lüders was, of course, published first.

While in Birmingham we married, on 01.05.1954; I was also in the Walkinshaw group, and, like John, was also a vegetarian, in my case from birth.

In the autumn of 1954 John returned to Harwell, to the group of the late T. H. R. Skyrme. There he continued research on field theory. He received a Ph.D. from Birmingham in 1956, the second half of his thesis being 'Functional Methods in Field Theory'. There were papers in this period on nuclear physics, many-body problems, and nuclear optical models. Collaborators included T. H. R. Skyrme, J. L. Perring, R. J. Blin-Stoyle, and Prof. E. J. Squires.

CERN (and SLAC)

In November 1960 the Bells moved to CERN, Geneva, John going to the Theory Division there. Apart from a year at SLAC (Stanford Linear Accelerator Center), from November 1963 until December 1964, John remained there until his sudden death in October 1990. He travelled a lot, however, on short trips giving talks. Many papers on a variety of subjects appeared – neutrino reactions, weak interactions and current algebra, vector boson production.

Among his collaborators were Profs. M. Veltman, J. Steinberger, R. Jackiw (see this book) and H. Ruegg.

His interest in accelerators was renewed in the eighties with some theoretical papers, with M. Bell, relating to the ICE (Initial Cooling Experiment). Indeed he was responsible for the name 'ICE'. The experiment formed part of the preparation for the antiproton accumulator, or collector. There were also papers on radiation damping and beamstrahlung, and papers with J. M. Leinaas and R. J. Hughes on the Unruh effect and quantum fluctuations in storage rings (see J. M. Leinaas in this volume).

We have kept the foundations of quantum mechanics comments to the end. From his time as a student he had been dissatisfied with the subject (see, e.g., Quantum Profiles, by Jeremy Bernstein, Princeton University Press, 1991). In the early fifties the papers on hidden variables by David Bohm appeared, and he was extremely stimulated by these. He had many heated arguments at that time with Franz Mandl, who had the advantage of being able to read the Von Neumann proof of the impossibility of a hidden variable interpretation of quantum mechanics (it was in German). The problem simmered in his mind, until, while at SLAC, he had time for reflection. There he wrote 'On the Problem of Hidden Variables' which disposed of the Von Neumann proof. A second paper, 'On the Einstein, Podolsky and Rosen Paradox', showed that any such interpretation has to involve non-local quantum logic. The violation of the associated Bell's inequalities was demonstrated, after earlier less refined experiments, by Alain Aspect, a French physicist. Many other papers on the subject appeared through the years (see the paper by Prof. K. Gottfried, in this volume).

A few personal notes

Although John was very practical he always studied the literature on any subject in which he was interested. As a boy he learned to swim after reading 'Every Boy and Girl a Swimmer', and to dance after reading the Victor Sylvester book. On arriving in Switzerland he found a French skiing book which was full of dynamics and discussion of conservation of angular momentum. He declared that this made the ski instructor's lessons much clearer. In later years his Celtic temperament calmed down as he became older, but he always had difficulty in relaxing. All his life he suffered from migraine, perhaps due to an overconscientious nature. He was always kind and generous, although easily irritated by 'woolly thinking'. All his life he supported, among other charities, especially World Population Control and the Replacement of Animals in Medical Research.

Mary Bell

Honours

1982– Reality Foundation Prize (shared with J. F. Clauser).
1987– Elected Honorary Foreign Member of the American Academy of Arts and Sciences.
1988– Dirac Medal of the Institute of Physics (London).
1988– D.Sc. h.c. Queen's University, Belfast.
1988– Sc.D. h.c. Trinity College, Dublin.
1989– Dannie Heinemann Prize for Mathematical Physics, American Physical Society.
1989– Hughes Medal, Royal Society, London.

1 On Bell non-locality without probabilities: some curious geometry

ROGER PENROSE

In 1966, John Bell [1] showed how Gleason's 1957 [2] theorem can be used to demonstrate the incompatibility of the predictions of quantum theory with 'non-contextual' hidden variable models. Later, Kochen and Specker [3] independently found a set of 117 (unoriented) spatial directions that exhibited this incompatibility in a finite explicit way. Such configurations have been used [4, 5, 6] as part of an EPR system, to show that the non-contextual assumption can be replaced by one of locality. This, like results obtained recently by Greenberger, Horne, Zeilinger (GHZ) [7] and others illustrates a conflict between quantum mechanics and locality that shows up in yes/no constraints on the results of certain idealized experiments, no probabilities being involved. Kochen and Specker's original set of 117 directions, for a 3-state (spin 1) system, has recently been reduced to 33 by Peres [8] (and to 31 by Conway and Kochen). Peres has also exhibited a set of 24 Hilbert-space directions, with similar properties, for a 4-state system, these being the common eigenstates of sets of commuting operators among a set of nine found by Peres [9] (and Mermin [10]). In this chapter, I show how Peres's set of 33 directions can be directly visualized in terms of a geometrical configuration (three interpenetrating cubes) that appears in the Escher print 'Waterfall'. Using the Majorana description of general spin states, I also exhibit a quite different set of 33 idealized measurements that can be performed on a spin 1 system. These measurements are specified in terms of an explicit set of 18 oriented directions in space. The configuration involved in Peres's set of 24 Hilbert-space directions can be understood in terms of a 4-dimensional regular polytope known as the '24-cell', and they are, in principle, ideally suited to providing an EPR-type of GHZ non-locality without probabilities. Unfortunately, if each 4-state system is taken to be a spin 3/2 particle, no simple spatial geometrical description of the needed measurements seems to emerge. Instead, I provide an alternative configuration for spin 3/2, based on a regular dodecahedron, in which only 20 oriented directions are explicitly used.

1. Introduction

In this chapter, as a homage to John Bell and his foundational contributions to physics, I am pointing out some curious geometry that illustrates an aspect of his profound insights into the structure of quantum mechanics and its inconsistency with common sense. I imagine that the configurations that I shall describe might well have amused him.

In 1957 Gleason [2] proved a theorem concerning Hilbert spaces of dimension three or more. The importance of this result in demonstrating the inconsistency of the predictions of quantum theory with certain types of hidden variable theory (those now known as 'non-contextual') was pointed out by Bell [1]; see Brown [11] for an up-to-date survey. These considerations apply, in particular, to a particle of spin 1, where one envisages the measurement of the *square* of the spin value (i.e. the square S_q^2 of the spin operator S_q) in various directions q. The result of such a measurement is always either 1 or 0, taking units such that $\hbar = 1$. Note that the measurement does not depend upon the orientation of the direction in question (i.e. the measurement for q is the same as that for $-q$). I shall refer to an *unoriented* direction as a *line*, so it is merely the line in question that is relevant to a measurement of this type. Now, one finds that S_p^2, S_q^2 and S_r^2 commute for any mutually perpendicular triad of lines $|p|$, $|q|$ and $|r|$, so that a simultaneous measurement of the squared spin can be made for these three lines. The rules of quantum mechanics assert that for any three mutually perpendicular lines, the resulting measurements must always be two 1s and one 0 (since $S_p^2 + S_q^2 + S_r^2 = 2$). In 1967, Kochen and Specker [3] exhibited an explicit set of 117 lines which contained sufficiently many perpendicular lines among them that they explicitly exhibited an inconsistency of these quantum rules with the 'classical' requirement that for any specified line there should be a definite value for the spin squared, preassigned by the system (see also [12]). In 1990, Peres was able to reduce this to a set of 33 lines [8, 13] (and Conway and Kochen found a way of reducing the number to 31 at about the same time). In Section 2, I shall point out a remarkable relationship between Peres's set of 33 lines and a configuration of three interpenetrating cubes that the Dutch artist M.C. Escher used as an ornament in his well-known lithograph 'Waterfall'. (The Conway–Kochen set of 31 is not so symmetrical.) In Section 6, I show how, with spin 3/2 in place of spin 1, there is a corresponding configuration based on a regular dodecahedron, this involving 20 explicit (now oriented) directions (and another set of 20 Hilbert-space directions that are not explicitly involved in the measurements).

For a classical 'hidden variable' model of this behaviour, we imagine that the spin 1 system's response to any choice of triad of mutually perpendicular lines is predetermined before the measurement is performed on it. Imagine 'colouring' these lines either black or white, where WHITE means 0 and BLACK means 1. Then in order to model this quantum-mechanical behaviour of the spin in a 'classical' (non-contextual hidden-variable) way, we must have the following two properties:

(1.1) if p and q are perpendicular lines, then they cannot both be WHITE;

(1.2) if p, q and r are mutually perpendicular lines, then they cannot all be BLACK.

There is actually an additional assumption involved – and this is what is involved in the qualification 'non-contextual' – namely that the performing of a measurement (which I shall denote by $[pqr]$) with respect to a triad pqr of mutually perpendicular lines assigns the same value to p as would the performing of the corresponding measurement $[pst]$, where pst is also mutually perpendicular. Another way of putting this is that $[pqr]$ is equivalent to performing a measurement with respect to each line p, q, r separately (denoted $[p]$, $[q]$, $[r]$, respectively) and the order in which these measurements is performed is irrelevant. That this is the case for measurements in standard quantum mechanics is a feature of the commuting of the three operators involved. But a hidden-variable theory need not in general satisfy this criterion. If we consider the measurement $[pqr]$ to be a succession of measurements $[p]$, $[q]$, $[r]$ in turn, say, then we must allow for the possibility that $[p]$ disturbs the system in some particular way, so that measurement $[q]$ sees a different hidden-variable state from that seen by $[p]$, and which can have been influenced by the result of $[p]$; similar remarks would apply to $[r]$. Thus, the succession of measurements $[p]$, $[q]$, $[r]$ might produce a different result from the succession $[q]$, $[p]$, $[r]$, or from $[q]$, $[r]$, $[p]$, for example. Neither the 117 lines of the Kochen–Specker configuration, nor the 33 of the Peres configuration, can be all coloured BLACK or WHITE consistently with (1.1) and (1.2). Therefore no non-contextual hidden variable theory can produce agreement with all the yes/no predictions of quantum theory.

In 1983, Heywood and Redhead [4] provided a way of removing this non-contextual assumption and replacing it by a locality assumption, where the single spin 1 system is now replaced by an Einstein–Podolsky–Rosen (EPR) pair (see also [5, 6]). I provide a similar procedure in Section 5. In my examples, the EPR pair will be a pair of spin 3/2 atoms (or particles). This provides us with a type of non-locality, similar to that obtained recently by Greenberger, Horne, and Zeilinger [7], Greenberger, Horne, Shimony and Zeilinger [14], Mermin [10], and Clifton, Redhead and Butterfield [15], although they used generalized EPR-type experiments involving three or four separated subsystems, rather than the two considered here. Probabilities are not involved in any of these examples, the conflict between quantum mechanics and locality showing up in yes/no constraints on the results of idealized experiments.

In Section 3, I shall show how some of these considerations can be extended by use of Majorana's [16] description of general spin states, according to which the general state for spin $n/2$ is represented in terms of n points on the sphere. This will give us some new geometrical insights into the configurations that are involved; and on the basis of this description, I shall give some new configurations, generalizing those of the Kochen–Specker type, in Sections 4 and 6.

2. The Peres set of 33 lines and Escher's Waterfall ornament

Peres [8] was able to reduce Kochen and Specker's set of 117 lines to 33, using coordinates:

$(1,0,0)$, $(0,1,0)$, $(0,0,1)$,

$(0,1,1)$, $(0,1,-1)$, $(1,0,1)$, $(-1,0,1)$, $(1,1,0)$, $(1,-1,0)$,

$(0,1,\sqrt{2})$, $(0,1,-\sqrt{2})$, $(\sqrt{2},0,1)$, $(-\sqrt{2},0,1)$, $(1,\sqrt{2},0)$, $(1,-\sqrt{2},0)$,

$(0,\sqrt{2},1)$, $(0,-\sqrt{2},1)$, $(1,0,\sqrt{2})$, $(1,0,-\sqrt{2})$, $(\sqrt{2},1,0)$, $(-\sqrt{2},1,0)$,

$(1,1,\sqrt{2})$, $(1,1,-\sqrt{2})$, $(\sqrt{2},1,1)$, $(-\sqrt{2},1,1)$, $(1,\sqrt{2},1)$, $(1,-\sqrt{2},1)$,

$(1,-1,\sqrt{2})$, $(1,-1,-\sqrt{2})$, $(\sqrt{2},1,-1)$, $(-\sqrt{2},1,-1)$, $(-1,\sqrt{2},1)$, $(-1,-\sqrt{2},1)$.

In each case, the line in question joins the origin to the point with the coordinates as given (so the family of real multiples of each given coordinate triple describes the actual line). We can consider the above coordinate triples as providing us with a *graph*, i.e. as a set of points, called *vertices*, some pairs of which are connected with *edges*, the connections being without orientation. Each vertex is given by one of the coordinate triples, and it represents a line of the above configuration. Vertices connected with an edge represent perpendicular lines (i.e. pairs of coordinate triples whose scalar product vanishes). Now, we imagine each vertex to be coloured either BLACK or WHITE – representing the respective values 1 and 0 for a squared spin value along the line in question. The requirements for a non-contextual hidden variable description, as given above are: (2.1) no pair of connected vertices may both be WHITE; and (2.2) no triangle (triple of mutually connected vertices) can have all its vertices BLACK (see Fig. 1.1). Peres's result is that there is no way of colouring the 33 vertices of his graph consistently with these rules, so the impossibility of a non-contextual hidden variable model, consistent with quantum-mechanical predictions is thereby demonstrated.

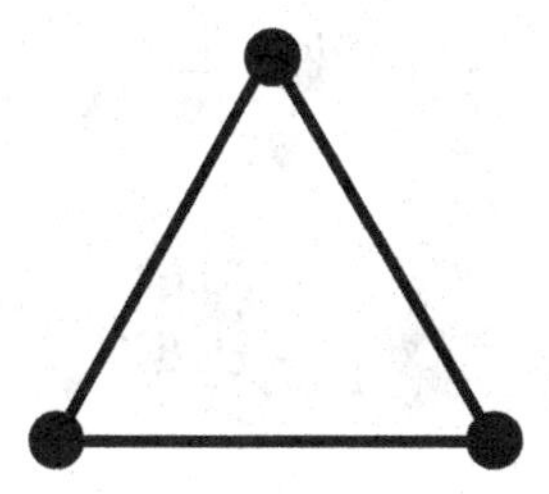

FIGURE 1.1 Forbidden colour configurations in the graph of squared spin measurements (WHITE = 1, BLACK = 0), for spin 1, where joined vertices represent perpendicular directions: no two connected vertices may both be WHITE; no three mutually connected vertices may be all BLACK. The graphs of neither the Kochen–Specker configuration (117 vertices) nor the Peres configuration (33 vertices) can be coloured consistently according to these rules.

FIGURE 1.2 M. C. Escher's lithograph 'Waterfall'. The ornament on the left-hand tower is a configuration of three interpenetrating cubes. The lines through opposite vertices, opposite edge mid-points, and opposite face mid-points of each cube give the Peres set of 33 directions.

It is not particularly difficult to verify this impossibility – if we take advantage of the evident symmetry of the Peres configuration under permutation and reflection of coordinate axes. I shall not go into the details of this here, but merely point out a remarkable relationship between the Peres configuration and one of the two ornaments on top of the towers appearing in the well-known print 'Waterfall' by the Dutch artist M.C. Escher (Fig. 1.2; left-hand tower). The ornament represents three interpenetrating cubes, each obtained from the others by 90° rotations about lines joining mid-points of its opposite edges. For each individual cube, we imagine lines of three kinds:

(F) 3 lines, joining mid-points of opposite faces
(E) 6 lines, joining mid-points of opposite edges
(V) 4 lines, joining opposite vertices.

This gives us 33 lines in all, there being some duplication among the lines of types (F) and (E) for the three cubes, three of them (the coordinate axes, about which the 90° rotations take place) counting three times over, so we get 33 lines rather than 39. Explicitly, for one of the cubes, we have lines as follows:

(F) $(1,0,0),(0,1,1),(0,1,-1)$
(E) $(0,1,0),(0,0,1),(\sqrt{2},1,1),(-\sqrt{2},1,1),(\sqrt{2},1,-1),(-\sqrt{2},1,-1)$
(V) $(1,\sqrt{2},0),(1,-\sqrt{2},0),(1,0,\sqrt{2}),(1,0,-\sqrt{2})$;

the lines for the other cubes are obtained from these by cyclic permutation of the coordinate axes. All the various orthogonality relations are obtainable by simple geometrical considerations concerning the cubes.

3. The Majorana description of general spin states

Some additional flexibility and insights concerning such configurations are obtainable if we take advantage of Majorana's [16] representation of the general state of total spin $n/2$ in terms of a set of n unordered points on the sphere (see also [17]). This representation can be understood [18] in terms of the two-component spinor formalism [19]. The spin state, for a spin $n/2$ particle, can be described as a symmetric n-indexed spinor $\psi_{ABC...F}$. Each index refers to a two-dimensional space (spin-space), and because of the symmetry we have $n+1$ independent components, each of which is a complex number:

$$\Psi_0 = \psi_{000...0}, \Psi_1 = \psi_{100...0}, \Psi_2 = \psi_{110...0}, \ldots, \Psi_n = \psi_{111...1}.$$

Any spinor of this nature can be expressed as a symmetric product of 1-index spinors, uniquely up to reorderings and rescalings, according to its *canonical decomposition* (see [19]):

$$\psi_{ABC...F} = \alpha_{(A}\beta_B\chi_C \cdots \phi_{F)},$$

where the bracketed indices on the right are to be symmetrized over. This is just a restatement of the so-called 'fundamental theorem of algebra' which asserts that any complex polynomial in one variable can be factorized completely into linear factors. We see this by introducing a spinor ζ^A, with components $\zeta^0 = 1$, $\zeta^1 = z$, and find

$$\begin{aligned}\Psi(z) &= \psi_{ABC...F}\zeta^A\zeta^B\zeta^C \ldots \zeta^F \\ &= \Psi_0 + n\Psi_1 z + \tfrac{1}{2}n(n-1)\Psi_2 z^2 + \cdots + \Psi_n z^n \\ &= (\alpha_0 + \alpha_1 z)(\beta_0 + \beta_1 z)(\chi_0 + \chi_1 z) \ldots (\phi_0 + \phi_1 z) \\ &= (\alpha_A\zeta^A)(\beta_B\zeta^B)(\chi_C\zeta^C) \ldots (\phi_F\zeta^F).\end{aligned}$$

The result is obtained by equating coefficients. The roots

$$z = -\alpha_0/\alpha_1, z = -\beta_0/\beta_1, z = -\chi_0/\chi_1, \ldots, z = -\phi_0/\phi_1$$

give us n (unordered) points on the Riemann sphere (the complex plane projected stereographically to the sphere, so that the point $z = \infty$ is also represented - given when $\Psi_n = 0$). The set of points defined by $\psi_{ABC...F}$ will be denoted by $[\psi]$. These points provide Majorana's representation of a general spin $n/2$ state (although he obtained it in a different way).

We wish to examine the condition of orthogonality between states given by spinors $\psi_{ABC...F}$ and $\theta_{ABC...F}$. This is defined in terms of the scalar product

$$\langle\psi|\theta\rangle = \psi_{ABC...F}t^{AA'}t^{BB'}t^{CC'} \ldots t^{FF'}\bar{\theta}_{A'B'C'...F'},$$

where I am here using the 2-spinors for 4-dimensional space-time, according to which the taking of complex conjugates interchanges the original spin-space (for which unprimed indices are used) with a separate 'conjugate' spin space (for which primed indices are used, A' being a distinct index letter from A, etc.). This way of looking at things will have a significance for us shortly. The quantity $t^{AA'}$ is the spinor equivalent of a timelike vector t^a, representing the time-axis. For our purposes, we simply take the components of $t^{AA'}$ to be the unit matrix. Then we have

$$\langle\psi|\theta\rangle = \Psi_0\bar{\Theta}_0 + n\Psi_1\bar{\Theta}_1 + \tfrac{1}{2}n(n-1)\Psi_2\bar{\Theta}_2 + \cdots + \Psi_n\bar{\Theta}_n.$$

Let us write

$$\omega^{ABC...F} = t^{AA'}t^{BB'}t^{CC'}\ldots t^{FF'}\bar{\theta}_{A'B'C'...F'}$$

and

$$\omega_{ABC...F} = \omega^{PQR...U}\varepsilon_{PA}\varepsilon_{QB}\varepsilon_{RC}\ldots\varepsilon_{EF},$$

where the skew-symmetrical Levi–Civita spinor ε_{AB} has components $(0,1;-1,0)$. Then we have

$$\begin{aligned}
\omega_{111...1} &= \bar{\theta}_{0'0'0'...0'}, &\quad \text{i.e. } \Omega_n &= \bar{\Theta}_0\\
\omega_{011...1} &= -\bar{\theta}_{1'0'0'...0'}, &\quad \text{i.e. } \Omega_{n-1} &= -\bar{\Theta}_1,\\
\omega_{001...1} &= \bar{\theta}_{1'1'0'...0'}, &\quad \text{i.e. } \Omega_{n-2} &= \bar{\Theta}_2\\
&\ldots & &\ldots\\
\omega_{000...0} &= (-1)^n\bar{\theta}_{1'1'1'...1'}, &\quad \text{i.e. } \Omega_0 &= (-1)^n\bar{\Theta}_n.
\end{aligned}$$

The orthogonality condition $\langle\psi|\theta\rangle = 0$, between $\psi_{...}$ and $\theta_{...}$, can now be restated as a condition holding between spinors $\psi_{ABC...F}$ and $\omega_{ABC...F}$, namely $\{\psi|\omega\} = 0$, where

$$\begin{aligned}
\{\psi|\omega\} &= \psi_{ABC...F}\omega^{ABC...F}\\
&= \psi_{ABC...F}\omega_{PQR...U}\varepsilon^{AP}\varepsilon^{BQ}\varepsilon^{CR}\ldots\varepsilon^{FU}\\
&= \Psi_0\Omega_n - n\Psi_1\Omega_{n-1} + \tfrac{1}{2}n(n-1)\Psi_2\Omega_{n-2} - \cdots + (-1)^n\Psi_n\Theta_0\\
&= (-1)^n\{\omega|\psi\}\\
&= \langle\psi|\theta\rangle.
\end{aligned}$$

An advantage of working with $\{\psi|\omega\}$, rather than with $\langle\psi|\theta\rangle$, is that $\{\psi|\omega\}$ is independent of $t^{AA'}$. This has the implication that the vanishing of $\{\psi|\omega\}$ states a condition holding between the two n-tuples $[\psi]$ and $[\omega]$ that is invariant under *conformal* motions of the sphere (given by transformations of the form $\zeta \to (\alpha\zeta+\beta)/(\gamma\zeta+\delta)$). The vanishing of $\langle\psi|\theta\rangle$ provides a condition on $[\psi]$ and $[\theta]$ that is merely invariant under *rotations* of the sphere (given when the matrix of α, β, γ, δ in the above transformation is unitary). The condition $\{\psi|\omega\} = 0$ is referred to as *apolarity* between the n-tuples $[\psi]$ and $[\omega]$ [20], the condition $\langle\psi|\omega\rangle = 0$ representing a (Hermitian) orthogonality between these n-tuples. The relationship between the specific n-tuples $[\omega]$ and $[\theta]$, as related by the above expression, is that each is the *reflection* of the other in the centre of the sphere, so to pass from apolarity to orthogonality we reflect one or other of the n-tuples in the centre. (See [19] for a description of the underlying spinor geometry.) It is of course orthogonality that we shall need for our quantum-mechanical applications, but apolarity is an interesting geometrical relationship and it is often easier to work with.

Apolarity is a complicated geometrical condition in general, but we shall be needing only certain special cases of it here. I give a few examples that will more than cover what we shall need. Bear in mind that there can be multiplicities amongst the points of $[\psi]$, the total of all the multiplicities being n.

(3.1) If $[\omega]$ has just one n-fold point, then $[\psi]$ and $[\omega]$ are apolar if and only if this n-fold point is one of the points of $[\psi]$.

PROOF: We see this directly from the above, with $\omega^{ABC...F} = \zeta^A\zeta^B\zeta^C\ldots\zeta^F$.

The following can be viewed as a generalization of the 'if' part of (3.1).

(3.2) If $[\omega]$ has a p-fold point and $[\psi]$ has the same point as a q-fold point, where $p+q>n$, then $[\psi]$ and $[\omega]$ are apolar.

PROOF: It is readily seen that every term of $\psi_{ABC...F}\omega^{ABC...F}$, when the symmetrizers are expanded out, must contain a factor of the form $(\zeta_A\zeta^A)$, which vanishes.

(3.3) If n is odd and the points of $[\psi]$ coincide with those of $[\omega]$, with the same multiplicities, then $[\psi]$ and $[\omega]$ are apolar.

PROOF: We have $\{\psi|\omega\} = \psi_{ABC...F}\omega_{PQR...U}\varepsilon^{AP}\varepsilon^{BQ}\varepsilon^{CR}\ldots\varepsilon^{FU} = -\{\omega|\psi\}$ if n is odd, because there is an odd number of skew-symmetrical ε-factors. Hence $\{\psi|\psi\} = 0$.

(3.4) If $[\psi]$ has an $(n-1)$-fold point, then $[\psi]$ and $[\omega]$ are apolar if the centroid of the stereographic projection of $[\omega]$ from this $(n-1)$-fold point is the remaining point of $[\psi]$.

PROOF: Rotate the sphere so that the $(n-1)$-fold point of $[\psi]$ is ∞. Then we have $\Psi_2 = \Psi_3 = \cdots = \Psi_n = 0$, so $\{\psi|\omega\} = \Psi_0\Omega_n - n\Psi_1\Omega_{n-1}$. The remaining point of $[\psi]$ is given by the vanishing of $\Psi(z) = \Psi_0 + n\Psi_1 z$, i.e. by $z = -\Psi_0/(n\Psi_1)$. Now, the sum of the roots of $\Omega(z)$ is always $-n\Omega_{n-1}/\Omega_n$. This sum, divided by n (i.e. $-\Omega_{n-1}/\Omega_n$), is the point in the complex plane that is the centroid of the n points in that plane that arise as stereographic projection of $[\omega]$ from the point ∞ on the Riemann sphere. This centroid is indeed $-\Psi_0/(n\Psi_1)$ if $\{\psi|\omega] = 0$, since then $\Psi_0\Omega_n = n\Psi_1\Omega_{n-1}$.

As a simple example of the application of (3.2), we can check that the $n+1$ eigenstates of the m-value of spin in some particular direction must indeed be mutually orthogonal. These $n+1$ states are easily recognizable from the fact that they are geometrically determined by the direction in question. That direction fixes a point P on the sphere, and also its antipodal point Q. The relevant n-tuples of points are: P as an n-fold point, P as an $(n-1)$-fold point and Q as a simple point, P as an $(n-2)$-fold point and Q as a double point, ..., Q as an n-fold point. (These give the respective spin-component eigenvalues $n/2$, $(n-2)/2$, $(n-4)/2, \ldots, -n/2$.) To ascertain the orthogonality between any two of these states, we first reflect one of them in the centre of the sphere so that its P becomes Q and its Q becomes P. Unless the two states were initially the same, this reflected state must be apolar to the unreflected one, by (3.2), because the sum of the multiplicities either at P or at Q must exceed n.

We can also use (3.1) to obtain an operational interpretation of the different Majorana points $[\psi]$, for a given state $\psi_{\ldots}$ [17, 18]. Consider a measurement of the m-value of the spin in the direction defined by one of these points. If it is a simple point, then the state which is n-fold in that direction is apolar to $[\psi]$. Therefore the state which is n-fold in the opposite direction (the one with eigenvalue $-n/2$) is orthogonal to $[\psi]$, so a measurement of the spin in a Majorana direction is characterized by the fact that it cannot yield the value $m = -n/2$. If the Majorana point were double, then the measurement could not yield $-n/2$ or $-(n-2)/2$; if triple then it could not yield $-n/2, -(n-2)/2, -(n-4)/2$; and so on. This all follows from (3.2).

4. Another set of 33 states for spin 1

We can use the Majorana description to construct various systems of states with orthogonality relations between them, some of which provide inconsistencies between non-contextual hidden variable models and the predictions of quantum theory. I shall give here a set of 33 such states for a system of spin 1.

These are essentially different from the system given by Peres, but the Peres set (and the Kochen–Specker set) can also be described in such terms. For this, we associate the measurement of the squared spin with a projection operator given by the state whose Majorana description is a pair of antipodal points along the line q in question. The squared spin S_q^2 itself is actually the projection *orthogonal* to this state, and I shall require the complementary projection $\mathbf{I} - S_q^2$, which projects *into* this Hilbert-space direction. Then the eigenvalue 1 is obtained if the system is in this state (WHITE), and 0 is obtained if the system is orthogonal to this state (BLACK). The projection into the state determined by the line q must be orthogonal to the projection determined by another line p whenever p and q are perpendicular (since then $\mathbf{I} - S_p^2$ and $\mathbf{I} - S_q^2$ commute – because S_p^2 and S_q^2 commute).

Let us see, in terms of the above geometry, why this must be the case. In fact, it is not hard to obtain the general condition of orthogonality for spin 1. For this, we need the apolarity condition between two doublets of points on the sphere: (A, B) and (C, D). We can see what this is from (3.4). Since $n = 2$, the $(n - 1)$-multiplicity condition is automatically satisfied for each point. Let us choose D as the point from which we project – 'stereographically', which means that we must project to a plane perpendicular to the line OD, where O is the centre of the sphere. Then, for apolarity, the projection of C must be the mid-point ('centroid') of the projections of A and B. This is known as the *harmonic* condition between the pairs (A, B) and (C, D). Another way of saying this is that A, B, C, D must all lie on the same plane (i.e. on a circle) and the straight line joining the points A, B meets the intersection of the tangent planes at C and at D. This intersection might possibly be at infinity, which means that the line AB is parallel to both tangent planes (though the two tangent planes need not be parallel to each other). The harmonic condition is symmetrical with respect to the interchange of (A, B) with (C, D), so it can be stated equally well with the roles of (A, B) and (C, D) reversed. Another way of stating the harmonic condition is that the distances between the points satisfy $AD \times BC = AC \times DB = AB \times CD/2$ (from which it follows that A, B, C, D all lie on a circle – a condition that can be used in place of one of the equalities if desired). The orthogonality condition between general spin 1 states is now easily obtained: the pair (A, B) is orthogonal to (E, F) if and only if (A, B) is harmonic with respect to the pair of antipodal points to (E,F). This is equivalent, of course, to (E,F) being apolar to the pair of antipodal points to (A,B).

Note that, by (3.1), the repeated point (A,A) is harmonic with respect to (C,D) if and only if either $C=A$ or $D=A$. Thus, (A,A) is orthogonal to any pair of points one of which is antipodal to A. Another simple example of harmonic points occurs when the two pairs are the pairs of opposite vertices of a *square* – or, more generally, pairs of opposite vertices of a *trapezoid*, so (A,B) and (C,D) are harmonic whenever A, B, C, D are coplanar, with distances $AC=BC$ and $AD=BD$. From these considerations we see, for example, that (P,Q), (R,S), (T,U) form an orthogonal triad, where P and Q are antipodal points, the lines RS and TU meeting PQ perpendicularly, the distance of RS from P being equal to the distance of TU from Q, where RS and TU are also perpendicular to each other.

The above considerations are sufficient to establish all the orthogonality relations that are required for the following set of 33 states. Consider a cube, and mark all the mid-points of the edges (*edge* points). Our sphere is chosen to pass through all these marked points (so it will be concentric with the cube at O) and, in addition, we mark the intersections of the cubic axes (through O) with the sphere (*face* points). The states are of four different types, given by Majorana descriptions:

(4.1) Three pairs of antipodal face points
(4.2) Six pairs of antipodal edge points
(4.3) 12 double edge points
(4.4) 12 pairs of edge points that are opposite across a face.

We have numerous orthogonality relations between these states. The three states in (4.1) are mutually orthogonal; each state in (4.1) forms a mutually orthogonal triad with a couple of states in (4.2) and with two different couples of states in (4.4); each state in (4.3) forms a mutually orthogonal triad with its opposite state in (4.3) and one of the states in (4.2); in addition, each state in (4.3) is orthogonal to two states in (4.4).

I shall leave it as an exercise for the reader to verify that it is not possible to 'colour' these states BLACK or WHITE in accordance with the rules (2.1), (2.2) (equivalent to (1.1), (1.2)): no two orthogonal states shall be WHITE; no orthogonal triad of states shall be entirely BLACK. This impossibility provides a new (but somewhat involved) argument showing the incompatibility between non-contextual hidden variable theories and the predictions of quantum mechanics. The measurements involved in the various projections to, in particular, the states of type (4.4) might be very hard to perform, however.

It should be pointed out that the configuration of 33 states in the construction given above is quite different from Peres's set of 33. If we form the graph of the states, as described in Section 2, then we get something topologically distinct, in the two cases, so there can certainly be no Hilbert-space rotation of one into the other.

5. Peres's 24 states for a 4-dimensional Hilbert space

Peres [13] has also given a much simpler set of states for the case of a 4-dimensional Hilbert space. These can be assigned coordinates

(A_1) $(1,0,0,0),(0,1,0,0),(0,0,1,0),(0,0,0,1)$
(A_2) $(-1,1,1,1),(1,-1,1,1),(1,1,-1,1),(1,1,1,-1)$
(A_3) $(1,1,1,1),(1,1,-1,-1),(1,-1,1,-1),(1,-1,-1,1)$
(B_1) $(1,1,0,0),(1,-1,0,0),(0,0,1,1),(0,0,1,-1)$
(B_2) $(0,1,1,0),(0,1,-1,0),(1,0,0,1),(1,0,0,-1)$
(B_3) $(0,1,0,1),(0,1,0,-1),(1,0,1,0),(1,0,-1,0)$.

Each of the quadruples of states in each row constitutes a mutually orthogonal tetrad. No (A_i) state is orthogonal to any (A_j) state unless $i = j$; nor is any (B_i) state orthogonal to any (B_j) state unless $i = j$. For each i and for each j, there are two disjoint orthogonal tetrads, each having two members from (A_i) and two members from (B_j). The colouring rules for these states that would have to be satisfied by a non-contextual hidden variable theory consistent with quantum theory would be:

(5.1) no two orthogonal states can be both WHITE;
(5.2) no four mutually orthogonal states can be all BLACK.

As in Section 4, WHITE means that the system is found to be in that state and BLACK means that it is found to be orthogonal to that state (i.e. eigenvalues 1 and 0, respectively, for the operator of projection to that state). It is a fairly direct matter to verify that no BLACK/WHITE colouring of these 24 states satisfies these rules.

So far, the 24 measurements are given merely abstractly, as projections to various directions in 4-dimensional Hilbert space. We would like to relate these abstract directions to some ordinary spatial geometry. The simplest system that comes to mind, having a 4-dimensional Hilbert space, is a system of spin 3/2. The question arises: is there a symmetrical-looking configuration of triplets of directions in ordinary Euclidean 3-space (Majorana description) that provide a concrete realization of these abstract states, where the orthogonality relations are reasonably apparent from geometrical considerations like those of Section 4.

Since any 3-dimensional complex Hilbert space is the same, abstractly, as any other, there are certainly many realizations of Peres's set of 24 states in terms of Majorana triples. Indeed, it is not hard to construct these in terms of sets of solutions of cubic equations, but as things stand, I have not been able to find a configuration that is geometrically interesting or for which all the orthogonality relations are reasonably manifest. There is, however, such a symmetrical configuration that does come tantalizingly close to satisfying all the required conditions. Take a regular octahedron and consider 24 states all of whose Majorana descriptions are of the form (X, X, Y), where X is a vertex and where Y lies at 120° from X, as measured out from the centre, on a great circle through the vertex X and three others. Then we find that almost all the required orthogonality conditions are satisfied, but there is one geometrical arrangement of pairs of states $(X, X, Y), (U, U, V)$ for which it fails – though in this case it turns out that (X, X, Y) and (U, V, V) are actually orthogonal! In the hope that someone else might be able to make something of this, I give some of the orthogonal configurations of points that are relevant. Consider a regular dodecagon with its 12 vertices, ordered X_0, X_1, X_2, $X_3, \ldots, X_{11}$, on a great circle of the sphere. Then (X_0, X_0, X_4) is orthogonal to (X_6, X_6, X_2), (X_6, X_6, X_{10}), (X_0, X_0, X_8), (X_8, X_8, X_4), (X_3, X_3, X_7), (X_9, X_9, X_1), and to any other state of the form (X_6, X_6, Y).

As a configuration in Euclidean 4-space, the cartesian coordinates (A_i) and (B_i), together with their reflections in the origin, form a configuration known previously to geometers: the 12 (B_i) points – together with their reflections – constitute the 24 vertices of a regular polytope known as a '24-cell', made up of 24 regular octahedra, 8 of these octahedra coming together at each vertex (see [21]). The centres of these octahedra form an exactly similar configuration, these points being given by the (A_1) points, and by the (A_2) and (A_3) points at half distance out from the origin (and by the reflections of all these points in the origin).

This configuration is closely related to a set of nine operators, introduced by Peres [9], partially following on from a suggestion of Mermin – originally for a pair of spin 1/2 particles rather than for a single state of spin 3/2. The following set of 9 operators is equivalent to those:

$$\sigma_x \otimes \sigma_x, \quad \sigma_x \otimes \sigma_y, \quad \sigma_x \otimes \sigma_z,$$

$$\sigma_y \otimes \sigma_x, \quad \sigma_y \otimes \sigma_y, \quad \sigma_y \otimes \sigma_z,$$

$$\sigma_z \otimes \sigma_x, \quad \sigma_z \otimes \sigma_y, \quad \sigma_z \otimes \sigma_z,$$

the first σ (Pauli matrix) referring to the left-hand particle and the second σ referring to the right-hand one, x, y, z being a mutually perpendicular triad in space. Each selection of three of these operators, for which there are no two in the same row and no two in the same column, form a mutually commuting triple. The various sets of common eigenvectors for these triples give Peres's 24 states, where the basis is provided by (suitable multiples of) the four states: up/up, up/down, down/up, down/down. The operators themselves are doubly degenerate, having four of the 24 states with eigenvalue $+1/4$ and four with eigenvalue $-1/4$. However, they are not operators that could be easily measured. One has to envisage some kind of 'entangled' measurement on both particles together which does not separate the eigenvector associated with the eigenvalue pair $(+1/2,\ +1/2)$ from that associated with $(-1/2,\ -1/2)$, and which does not separate the one associated with $(+1/2,\ -1/2)$ from that associated with $(-1/2,\ +1/2)$.

These kinds of consideration apart, the Peres system of 24 states ought to provide an ideal arrangement for an Einstein–Podolsky–Rosen type of thought experiment in which the assumption of non-contextuality is replaced by one of locality. We consider two spin 3/2 atoms (or particles) emitted from a system, one to the left and the other to the right, so that their combined state of spin is a singlet state (spin 0). We propose to measure the left-hand atom according to one of three alternative measurements $[A_1]$, $[A_2]$ or $[A_3]$, and the right-hand one according to one of $[B_1]$, $[B_2]$ or $[B_3]$. Each measurement $[A_i]$ is to be non-degenerate, with eigenvectors given by the tetrad (A_i), above, according to some chosen orthonormal basis for spin 3/2 states; likewise, each measurement $[B_i]$ is to be similar, with eigenvectors given by the tetrad (B_i), referred to the orthonormal basis for spin 3/2 states which is that obtained from the previous one by space-reflection in the centre.

Now, suppose that we had a local hidden-variable model that could reproduce the quantum-mechanical results for this set-up. Let us first imagine that the left-hand measurement is $[A_1]$. The result might turn out to be, say $(1,0,0,0)$, but if so, then the left-hand atom had *no choice* but to have been committed to produce this value ahead of time. Let us see why this must be the case. First, the left-hand atom could have no 'knowledge' – for a *local* hidden-variable model – that a measurement $[B_1]$ *might* not be performed on the right-hand atom, in which case the result of $[B_1]$ would have to be either $(1,1,0,0)$ or $(1,-1,0,0)$, the other two being orthogonal to $(1,0,0,0)$. (Recall that the combined state must have spin 0 and that the $[B_i]$ measurements are described, relative to the description of the $[A_i]$ measurements, by space-reflection in the centre.) Therefore the measurement $[A_1]$ on the left-hand atom *had* to have been prepared to produce *either* $(1,0,0,0)$ or $(0,1,0,0)$, the other two being orthogonal to both of $(1,1,0,0)$ and $(1,-1,0,0)$. Furthermore, the left-hand atom could have no knowledge that a measurement $[B_2]$ might not be performed on the right-hand atom, so it similarly follows that the measurement $[A_1]$ on the left-hand atom had to have been prepared to produce either $(1,0,0,0)$ or $(0,0,0,1)$. Finally, the left-hand atom could have no knowledge that $[B_3]$ might not be performed on the right-hand atom, whence the measurement $[A_1]$ on the left-hand atom had to have been prepared to produce either $(1,0,0,0)$ or $(0,0,1,0)$. Thus, it indeed follows that the left-hand particle must have been committed to produce the value $(1,0,0,0)$ ahead of time.

Exactly the corresponding conclusion would follow *whatever* the result of $[A_1]$. Similarly, the left-hand atom would have to have its answers prepared if $[A_2]$ is performed on it or if $[A_3]$ is performed on it. The same holds with regard to the three possible measurements $[B_1], [B_2], [B_3]$ that might be performed on the right-hand particle.

This, in effect, removes any role for contextuality, and it is merely a matter of checking through the possibilities to verify that no assignment of values for all these possible measurements can be made consistently with the rules (5.1), (5.2). As a final remark in this section, it should be pointed out that the 4-state systems on the left and right need not be spin 3/2 atoms, but any other pair of 4-state objects would do – say, two pairs of spin 1/2 particles, as in the original Greenberger, Horne and Zeilinger [7] example.

6. An EPR spin 3/2 pair, with directions based on a dodecahedron

The main problems with the previous example are the lack of a clear geometric picture in ordinary space and the inherent difficulty of making measurements of the types that would be needed – either non-local or projections to an awkward Majorana configuration. In this final section, I provide a different example, in which these difficulties are effectively removed. (They would also be removed if we were to adapt Peres's configuration of 33 lines to an appropriate EPR context, or the original Kochen–Specker set of 117, as done by Heywood and Redhead [4], but thc cxample that I give is simpler, and it has its own special geometric appeal.)

The spin 3/2 states that I shall be considering will have Majorana descriptions of the form (X, X, Y), where X and Y are antipodal points on the sphere; that is to say, they are eigenstates, with eigenvalue 1/2, of a spin measurement in the direction OX, from the centre O of the sphere. The essential fact that we shall need to know is that two such states (X, X, Y) and (U, U, V) are orthogonal if X and U are two adjacent vertices of a cube inscribed in the sphere (U and V being antipodal vertices, like X and Y). To see this, consider Fig. 1.3. The rectangle $XUYV$ has sides in ratio $1 : \sqrt{2}$, from which it follows (by use of similar triangles) that the projection U' of U from the point X to the line through Y perpendicular to XY (tangent line at Y to the circle $XUYV$) is twice as far from Y as is the corresponding projection V' of V. Counting V with multiplicity 2, it follows from (3.4) that (X, X, Y) and (V, V, U) are apolar, whence (X, X, Y) and (U, U, V) are indeed orthogonal. Note that the angle α between the directions OX and OU is defined by $\tan(\alpha/2) = 1/\sqrt{2}$, i.e. $\alpha = \cos^{-1}(1/3) \approx 70.53°$.

The 20-state configuration that I wish to consider consists of states of the form (X, X, Y), where the points X are the various vertices of a *regular dodecahedron* (Y being antipodal to X in each case). We are considering measurements of spin in these 20 directions, each of these measurements being a projection to the state with eigenvalue $+1/2$. If the atom is found to have the spin value $m = +1/2$ in that direction, then we colour the vertex of the dodecahedron in that direction WHITE (i.e. 'yes'); if it is found to be orthogonal to that direction, then we colour the corresponding vertex BLACK ('no'). We consider such measurements being performed on an EPR pair of spin 3/2 atoms emitted

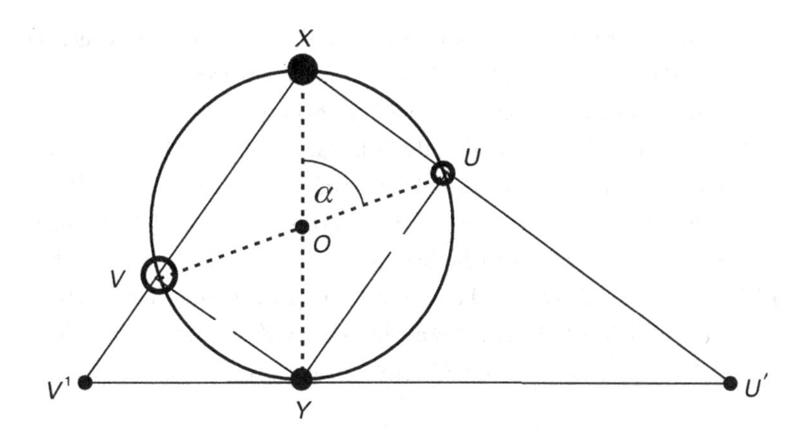

FIGURE 1.3 The two triples (X, X, Y) and (V, V, U) are apolar when $XUYV$ forms a rectangle for which the distance XV is $\sqrt{2}$ times the distance XU. This follows from (3.4), since it can be seen by simple use of similar triangles that if we project U and V (stereographically) from X to the respective points U', V' on the tangent at Y to the circle $XUYV$, then the distance YU' is twice the distance $V'Y$. The pair of points U, X subtend the angle α at the centre O of the circle.

originally in a combined singlet state, just as in Section 5. In my descriptions, the colouring of the dodecahedron will refer to the *left*-hand atom; to obtain the colouring for the right-hand atom, we should have to reflect the colouring for the left-hand atom in the centre O. I shall show that for a local realistic model, consistent with the predictions of quantum theory, the colouring for the vertices of the (left-hand) dodecahedron must satisfy:

(6.1) the result of such a spin measurement in any vertex direction must have been decided in advance – i.e. each vertex must be coloured definitely either WHITE or BLACK;

(6.2) no two next-to-adjacent vertices can be both WHITE;

(6.3) the six vertices adjacent to any pair of antipodal vertices cannot all be BLACK.

By 'next-to-adjacent' vertices, in (6.2), I mean non-adjacent vertices that belong to the same pentagonal face. There is also another rule, included only for completeness, that will not be needed here:

(6.4) no two antipodal vertices can be both WHITE.

I then show that there is no colouring of all the dodecahedron's vertices consistent with the rules (6.1), (6.2), (6.3), so no local realistic model consistent with the yes/no predictions of quantum theory is possible.

We first note that the angle subtended at the centre O by a pair of next-to-adjacent vertices of the dodecahedron is the same angle $\alpha = \cos^{-1}(1/3)$ that is subtended at the centre by two adjacent vertices of a cube. In fact, the entire set of vertices for the cube can be located as a subset of vertices of the dodecahedron, where adjacent cube vertices are next-to-adjacent vertices of the dodecahedron (see Fig. 1.4). Hence, by what was said above, the states determined by a pair of next-to-adjacent vertices are orthogonal. Moreover, the

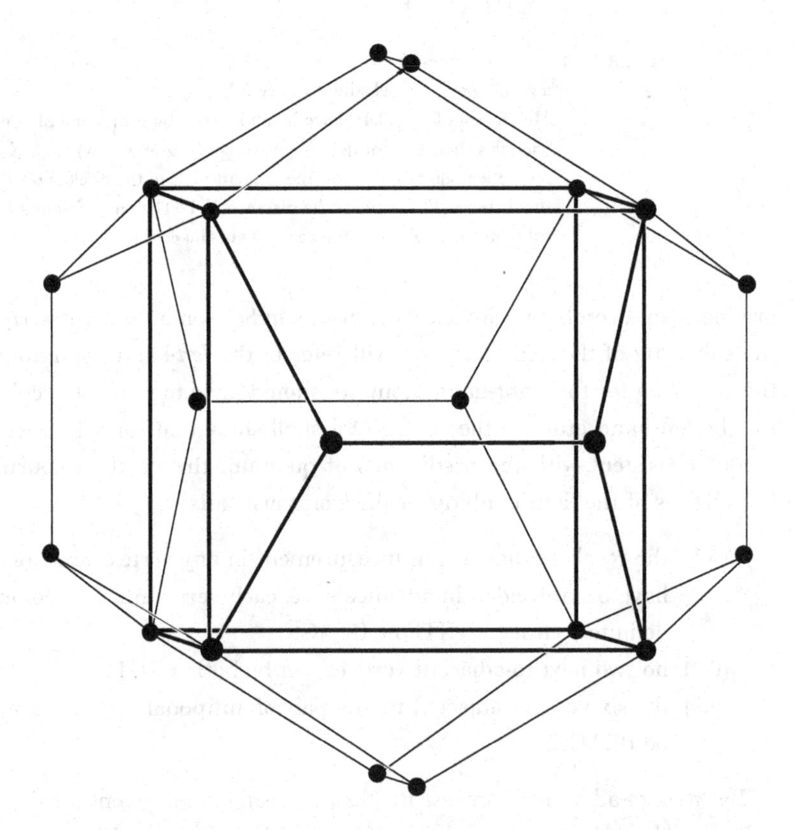

FIGURE 1.4 A cube can be inscribed in a regular dodecahedron such that adjacent vertices of the cube are next-to-adjacent vertices of the dodecahedron. Each vertex of the dodecahedron lies on two such cubes, there being five cubes in all.

three vertices that are adjacent to a given vertex of the dodecahedron are all next-to-adjacent to each other and they therefore determine three states that are mutually orthogonal. There will be a fourth state that completes a tetrad of mutually orthogonal states with these three, but that is not one of the states that we have been considering up to this juncture – I shall refer to such a state as *implicit.* Its Majorana description is not of the form (X, X, Y), being instead a triple of distinct points symmetrically positioned about the axis through the original given vertex. It will not be necessary for us to locate these points precisely, but one fact about this state will be needed: it is orthogonal to the corresponding state that completes the tetrad of mutually orthogonal states determined by the three vertices adjacent to the *antipodal* point to the original chosen vertex. The reason for this is that the Majorana descriptions of these two implicit states must be antipodal to one another. (They are determined in a geometrical way by sets of points that are antipodal to each other.) That two antipodal triplets of points must represent states that are orthogonal to one another follows from (3.3) above.

The type of idealized measurement that I envisage, on a spin 3/2 atom, involves ascertaining whether or not the m-value of its spin is $+1/2$ in some given direction OX. This measurement must be such that if the value $m = +1/2$ is *not* obtained, then the phase relations between the states for the remaining m-values, $+3/2, -1/2, -3/2$, are not disturbed, so the measurement represents a genuine projection to the $m = +1/2$ eigenstate, with complete degeneracy for the remaining three eigenstates. If the value $m = +1/2$ *is* obtained, the answer to the measurement is 'yes', so X is then coloured WHITE; if this value is *not* obtained, we simply get 'no', there being no further information about the m-value from the measurement, and X is coloured BLACK. The measurement might be made by some modified form of Stern–Gerlach apparatus, perhaps by suspending the atom at rest in an electric field and then separating the four m-eigenstates with an inhomogeneous magnetic field, where the atom is assumed to have a magnetic moment aligned with its spin. The resulting position, corresponding to $m = +1/2$, is then examined, and if the atom is found *not* to be there the inhomogeneity of the magnetic field is appropriately reversed so as to bring the three remaining states $m = +3/2, -1/2, -3/2$ together, ready for the procedure to be repeated in another direction, this new direction making an angle α with the first. If, according to this second procedure, the atom is again not found to be in the spot appropriate for $m = +1/2$ in this new direction, then the magnetic field inhomogeneity would be reversed again and the whole procedure repeated one further time, in a new direction making an angle α with both the previous directions. For any triple of directions, each of which making an angle α with the two others, either there could be exactly one 'yes' among the three or all three could be 'no'. (As soon as the atom *is* found, in an $m = +1/2$ spot, the procedure is stopped.)

Let us now try to use measurements of this kind in an EPR-type experiment in order to derive the properties (6.1), (6.2), (6.3), above. We assume that we have two spin 3/2 atoms, having been emitted to the left and to the right in a state with zero total angular momentum. According to standard quantum mechanics, if a measurement on the atom on the left finds that atom to be in a particular state of spin, say a state with Majorana description (P, Q, R), then the corresponding measurement on the right (but reflected in the centre) must give (P', Q', R'), where P', Q' and R' are, respectively, the antipodal points of P, Q and R. Now let us assume that a measurement, as above, is made on the left-hand atom, corresponding to the direction X, where X is some vertex of the dodecahedron. Suppose, first, that the left-hand atom is indeed found to have $m = +1/2$ (i.e. 'yes'), then it follows from the locality assumption that the spin value could not have been anything other than this. We shall see that it would have to have been 'prepared in advance' according to any local hidden-variable theory, by hidden parameters in the atom and/or measuring apparatus. The reason is that the *right*-hand atom *might* be measured in the direction Y, where Y is antipodal to X, and if so, it would also have to provide the answer 'yes', but the left-hand atom and measuring device could not have 'known' that the right-hand atom might have been measured in this way, so its own answer 'yes' would indeed have had to have been determined by its own state and the state of the measuring apparatus ahead of time. Exactly the same consideration would apply had the answer been 'no' on the left, since this would have to be accompanied by 'no' also on the right for the antipodal measurement.

We must also establish the non-contextual nature of these measurements. For measurement of the type described above, this amounts to establishing that it makes no difference whether the spin measurement in some particular direction is the first, second, or third of the triple (each making an angle α with the others) that can be performed on the atom. It would then follow that the triple of measurements can be thought of as a single measurement, with four possible outcomes: 'yes' in one of the three chosen directions, or 'no' for all three – the last of which means that there is an implicit 'yes' for the unmeasured (implicit) state orthogonal to the explicitly measured three. The non-contextual nature of any one of the three explicit measurements would follow because any one of them could have been the first to be measured, in which case it is immaterial which directions are chosen for the second and third measurements. The fact that the ordering of the measurements is indeed immaterial, according to the predictions of ordinary quantum mechanics, is a consequence of the fact that the measurement corresponding to projection to $m = +1/2$ in some direction is *orthogonal* to that corresponding to a direction making an angle α to that direction. Experimentally, one would have to check that in fact the ordering does not matter by performing measurements on both the left-hand and right-hand atoms, using independent orderings on the two sides, and making sure that the results of the left-hand measurements always agree with the antipodal right-hand measurements whatever the orderings are on either side. This would provide an experimental verification of (6.1).

Moreover, (6.2) would also be verified by these measurements because if two next-to-adjacent vertices U and X are both WHITE for the left-hand atom (the angle between OX and OU being α), the measurement of the left-hand atom with respect to a triple that includes the OX direction, measuring OX first, would give 'yes' for X, and that of the right-hand atom with respect to the triple antipodal to that one, but for which the measurement in the direction OV, opposite to OU, is performed first, would also give 'yes' for V, which contradicts the predictions of quantum mechanics as described above.

To verify (6.3), we make sure that for any triple measurement on the left, and for the parallel triple of measurements on the right, it is never found that all six measurements give 'no'. That this should be the case, according to quantum mechanics, follows from the discussion given earlier: if the three on the left all give 'no', then the state ψ of the left-hand atom must actually be the *implicit* one for that triple, but ψ is orthogonal to the implicit state θ which is antipodal to it, so we would get 'no' for the measurement on the right-hand atom that projects to the state antipodal to θ (i.e. ψ) were we to attempt to measure it, which means that we must get a 'yes' for one of the directions of the triple on the right.

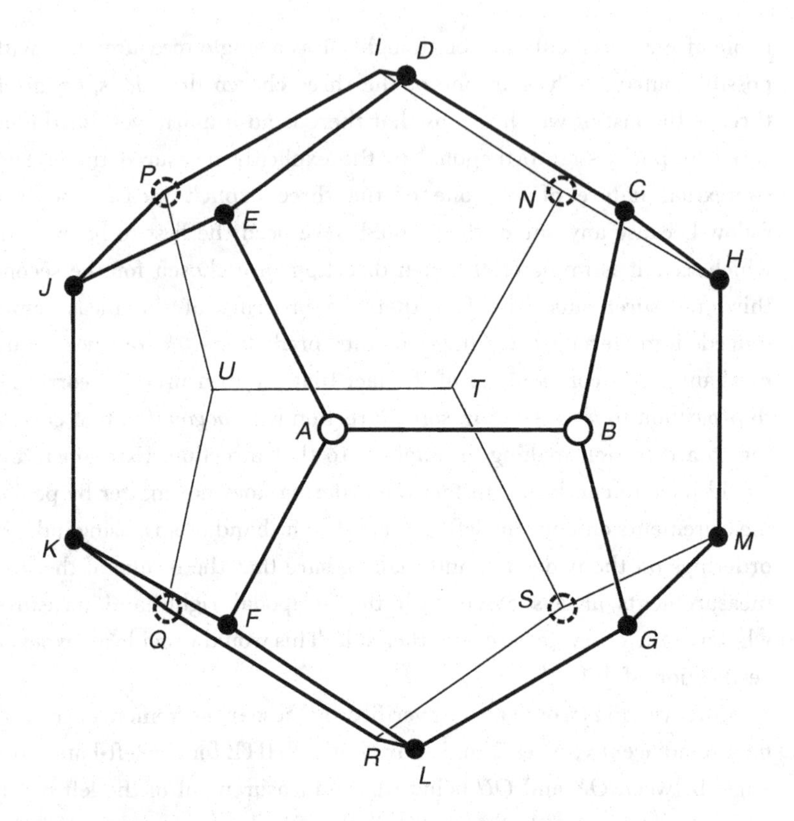

FIGURE 1.5 If adjacent vertices A, B of the dodecahedron are both coloured WHITE then, by (6.2), all the vertices C, D, E, J, K, F, L, G, M, H must be BLACK. By (6.3), the six vertices adjacent to J and M cannot all be BLACK, so at least one of P, S must be WHITE; similarly, at least one of N, Q must be WHITE. This contradicts (6.2), since P and S are each next-to-adjacent to each of N and Q.

Finally, we must check that no colouring of the dodecahedron in accordance with the rules (6.1), (6.2), (6.3) is possible. The following description refers to the labelling of the twenty vertices given in Fig. 1.5. We first note that not all the vertices can be BLACK, since that would violate (6.3). Without loss of generality, we can assume that A is WHITE. Now suppose that there is a vertex adjacent to A that is also WHITE. Without loss, we can assume that WHITE vertex to be B. Using (6.2), we can colour BLACK all the vertices that are next-to-adjacent to either A or B. We arrive at the configuration of colours depicted in Fig. 1.5. Next, we note that the triples K, E, P and G, H, S are respectively adjacent to the antipodal pair J, M so, by (6.3), P and S cannot both be BLACK; similarly, N and Q cannot both be BLACK. But this gives two non-adjacent WHITE vertices: a contradiction with (6.2). Thus, our assumption that there can be a WHITE vertex adjacent to A – or, indeed, adjacent to any other WHITE vertex – must be false: B, E, and F must all be BLACK. From (6.3) it then follows that there must be a WHITE vertex amongst U, S, N (since these two triples are adjacent to the antipodal pair A, T). See Fig. 1.6. Without loss, we can assume N to be WHITE. Then, by (6.2), S and P must be BLACK,

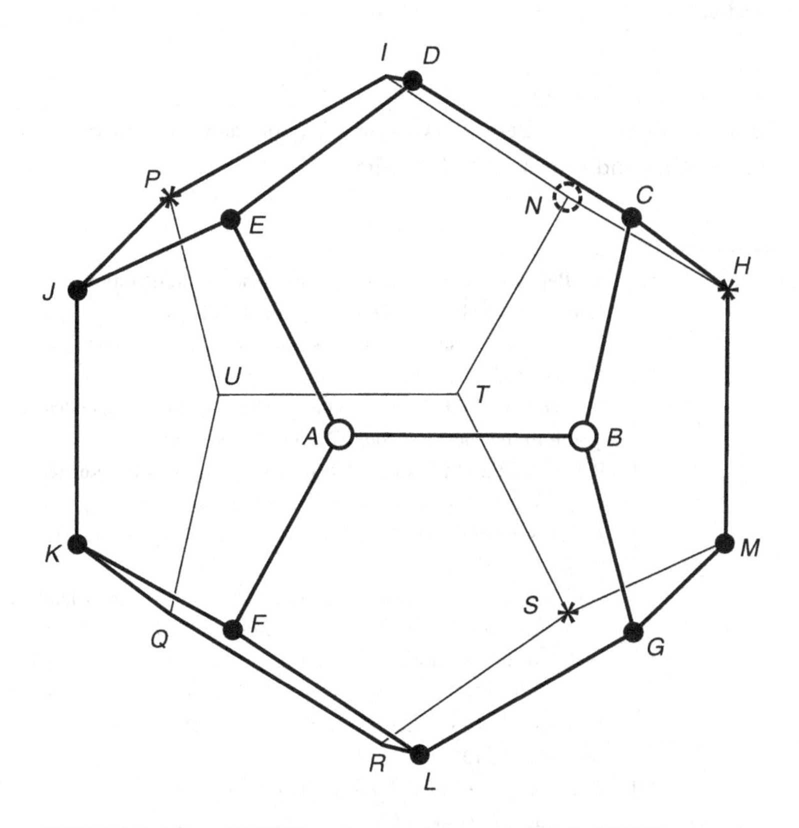

FIGURE 1.6 Adjacent WHITE vertices now being excluded, the WHITE vertex A must be surrounded by BLACK vertices $B, C, D, E, J, K, F, L, G$. By (6.3), the six vertices adjacent to A and T cannot all be BLACK, so at least one of N, U, S must be WHITE, and we can assume, without loss of generality that N is WHITE. Hence H, P, S must all be BLACK, by adjacency or next-to-adjacency with N; but this contradicts (6.3) since we now have six BLACK vertices adjacent to the antipodal pair J, M.

since they are next-to-adjacent to N; moreover, H must also be BLACK because it is adjacent to N, and we have ruled out adjacent WHITE vertices. The triples S, H, G and E, K, P are now all coloured BLACK (since G and K are already next-to-adjacent to A) which contradicts (6.3) because they are adjacent to the antipodal pair M, J.

This contradiction with the rules (6.1), (6.2), (6.3) provides us with a further argument, free of probabilities, of John Bell's very remarkable conclusion: no local realistic model can be consistent with the predictions of quantum mechanics.

ACKNOWLEDGEMENTS

I am grateful to Asher Peres, David Mermin, and especially Harvey Brown for enlightening and informative discussion.

REFERENCES

[1] J.S. Bell, On the problem of hidden variables in quantum theory, *Revs. Mod. Phys.* **38** (1966) 447–52. Reprinted, Bell [23].

[2] A. Gleason, Measures on the closed subspaces of a Hilbert space, *J. Math. Mech.* **6** (1957) 885–893.

[3] S. Kochen and E.P. Specker, The problem of hidden variables in quantum mechanics, *J. Math. Mech.* **17** (1967) 59–88.

[4] P. Heywood and M.L.G. Redhead, Nonlocality and the Kochen–Specker paradox, *Found. Phys.* **13** (1983) 481–499.

[5] A. Stairs, Quantum logic, realism, and value-definiteness, *Phil. Sci.* **50** (1983) 587–602.

[6] H.R. Brown and G. Svetlichny, Nonlocality and Gleason's lemma. Part I. Deterministic theories, *Found. Phys.* **20** (1990) 1379–87.

[7] D.M. Greenberg, M.A. Horne and A. Zeilinger, Going beyond Bell's theorem, in *Bell's Theorem, Quantum Theory, and Conceptions of the Universe*, ed. M. Kafatos (Kluwer Academic, Dordrecht, The Netherlands, 1989) pp. 73–76.

[8] A. Peres, Two simple proofs of the Kochen–Specker theorem, *J. Phys. A: Math. Gen.* **24** (1990a) L175.

[9] A. Peres, Incompatible results of quantum measurements, *Phys. Lett. A* (1990b) 107–8.

[10] N.D. Mermin, Simple unified form for the major no-hidden-variables theorems, *Phys. Rev. Lett.* **65** (1990) 3373.

[11] H.R. Brown, *Bell's other theorem and its connection with nonlocality.* Part I. Preprint; lecture at Cesena conference (1991).

[12] M.L.G. Redhead, *Incompleteness, Nonlocality, and Realism* (Clarendon Press, Oxford, 1987).

[13] A. Peres, Proc. NATO Adv. Res. Workshop on Quantum Chaos and Quantum Measurement, Copenhagen, 1991; (Kluwer, Dordrecht, 1992) 249.

[14] D.M. Greenberger, M.A. Horne, A. Shimony and A. Zeilinger, Bell's theorem without inequalities, *Am. J. Phys.* **58** (1990) 1131–43.

[15] R.K. Clifton, M.L.G. Redhead and J. Butterfield, Generalization of the Greenberger–Horne–Zeilinger algebraic proof of nonlocality, *Found. Phys.* **21** (1991) 149–84; errata *Found. Phys. Lett.* **4**.

[16] E. Majorana, Atomi orientati in campo magnetico variabile, *Nuovo Cimento* **9** (1932) 43–50.

[17] R. Penrose, *The Emperor's New Mind: Concerning Computers, Minds, and the Laws of Physics* (Oxford University Press, Oxford, 1989).

[18] R. Penrose, Newton, quantum theory and reality, in *300 Years of Gravity*, eds. S.W. Hawking and W. Israel (Cambridge University Press, Cambridge, 1987).

[19] R. Penrose and W. Rindler, *Spinors and Space-Time*, Vol. 1: *Two-Spinor Calculus and Relativistic Fields* (Cambridge University Press, Cambridge, 1984).

[20] J.H. Grace and A. Young, *The Algebra of Invariants* (Cambridge University Press, Cambridge, 1903).

[21] H.S.M. Coxeter, *Regular Polytopes* (Methuen, London, 1948).

[22] J.S. Bell, On the impossible pilot wave, *Found. Phys.* **12** (1982) 989–99. Reprinted, Bell [23].

[23] J.S. Bell, *Speakable and Unspeakable in Quantum Mechanics* (Cambridge University Press, Cambridge, 1987).

[24] É. Cartan, Sur les équations de la gravitation d'Einstein, *J. Math. Pures et Appl.* **1** (1922) 141–203 (p. 194).

[25] R. Penrose, *Shadows of the Mind* (Oxford University Press, Oxford, 1994), Sections 5.3, 5.18 and Appendices B and C.

[26] J. Zimba and R. Penrose, On Bell non-locality without probabilities: more curious geometry, *Stud. Hist. Phil. Sci.* **24** (1993) 697–720.

2 Reality in neutron interference experiments

HELMUT RAUCH

The wave–particle dualism becomes very obvious in matter wave interference experiments. Neutron interferometers based on wave front and amplitude division have been developed in the past. Most experiments have been performed with perfect crystal neutron interferometers which provide widely separated coherent beams enabling new experiments in the field of fundamental, nuclear and solid state physics. The difference between stochastic and deterministic absorption has been investigated in close connection to the quantum measurement problem. In the case of a deterministic absorption process the attenuation of the interference pattern is proportional to the beam attenuation, whereas in the case of stochastic absorption it is proportional to the square root of the attenuation. This permits the formulation of Bell-like inequalities which will be discussed in detail. The verification of the 4π-symmetry of spinors and of the quantum mechanical spin-superposition experiment on a macroscopic scale are typical examples of interferometry in spin space. These experiments were continued with two resonance coils in the beams where the results showed that coherence persists, even if an energy exchange between the neutron and the resonator system occurs with certainty. A quantum beat effect was observed when slightly different resonance frequencies were applied to both beams. In this case, the extremely high energy sensitivity of 2.7×10^{-19} eV was achieved. This effect can be interpreted as a magnetic Josephson-effect analogue. Phase echo, experiments with pulsed beams, and various postselection experiments show how interference phenomena can be made visible by a proper beam handling inside and behind the interferometer. All the results obtained until now are in agreement with the formalism of quantum mechanics but stimulate the discussion about the interpretation of this basic theory.

1. Introduction

Three different kinds of neutron interferometer have been tested in the past. The slit interferometer is based on wavefront division and provides long beam paths but only a very small beam separation [1, 2]. The perfect crystal interferometer [3, 4] is based on amplitude division and is now most frequently used due to its wide beam separation and its universal availability for fundamental, nuclear and solid state physics research. The interferometer based on grating diffraction is a recent development and has its main application for very slow neutrons [5]. A schematical comparison is shown in Fig. 2.1. The perfect crystal interferometer provides highest intensity and highest flexibility for beam handling.

In this chapter the development and the application of the perfect crystal interferometer are reviewed. The first successful test of such an interferometer happened in 1974 at our small 250 kW TRIGA-reactor in Wien [3] (Fig. 2.2).

The perfect crystal interferometer represents a macroscopic quantum device with characteristic dimensions of several centimetres. The basis for this kind of neutron interferometry is provided by the undisturbed arrangement of atoms in a monolithic perfect silicon crystal [3, 6]. An incident beam is split coherently at the first crystal plate, reflected at the middle plate and coherently superposed at the third plate (Fig. 2.1b). It follows immediately from general symmetry considerations that the wave functions in both beam paths, which compose the beam in the forward direction behind the interferometer, are equal ($\psi_0^{\mathrm{I}} = \psi_0^{\mathrm{II}}$), because they are transmitted-reflected-reflected (TRR) and reflected-reflected-transmitted (RRT), respectively. The system is based on Bragg diffraction from perfect crystals; therefore, the de Broglie wavelength of the neutrons is about 1.8 Å and their energy is about 0.025 eV.

Dynamical diffraction theory describes the diffraction process from the perfect crystal [7–10]. Inside the perfect crystal two wave fields are excited when the incident beam fulfils the Bragg condition, one of them having its nodes at the position of the atoms and the other in between them. Therefore, their wave vectors are slightly different ($k_1 - k_2 = 10^{-5} k_0$) and, due to mutual interference processes, a rather complicated interference pattern is built up, which changes substantially over a characteristic length Δ_0 – the so-called Pendellösung length, which is of the order of 50 μm for an ordinary silicon reflection. To preserve the interference properties over the length of

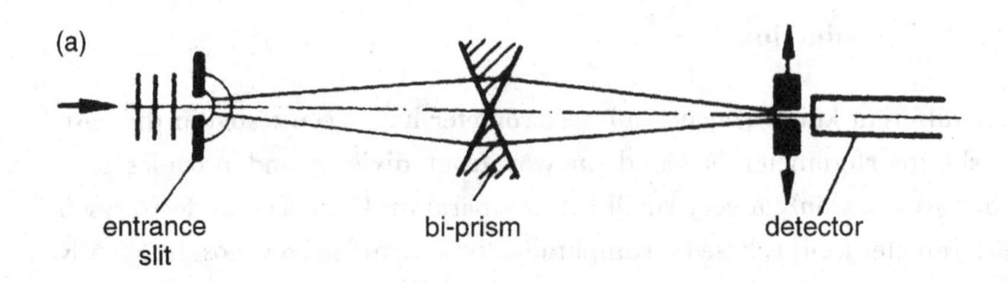

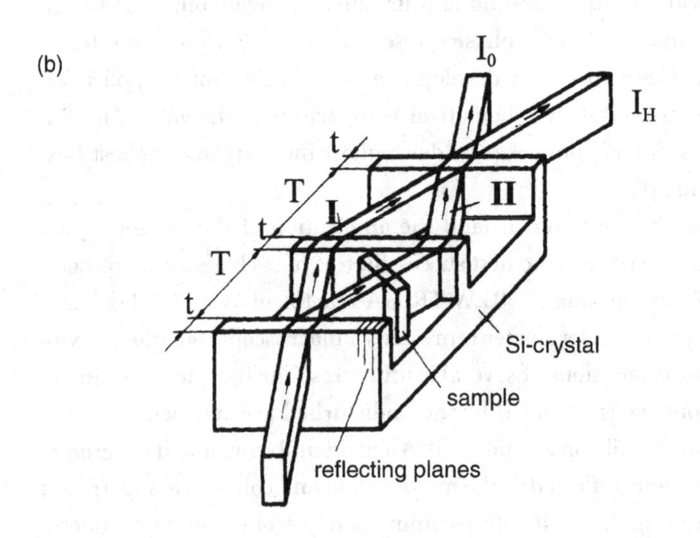

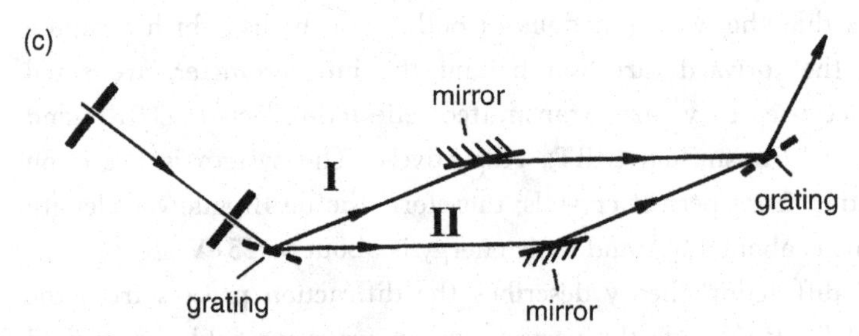

FIGURE 2.1 Scheme of (a) a slit, (b) a perfect crystal and (c) a grating interferometer.

the interferometer, the dimensions of the monolithic system have to be accurate on a scale comparable to this quantity. The whole interferometer crystal has to be placed on a stable goniometer table under conditions avoiding temperature gradients and vibrations to keep the lattice planes parallel within one lattice constant.

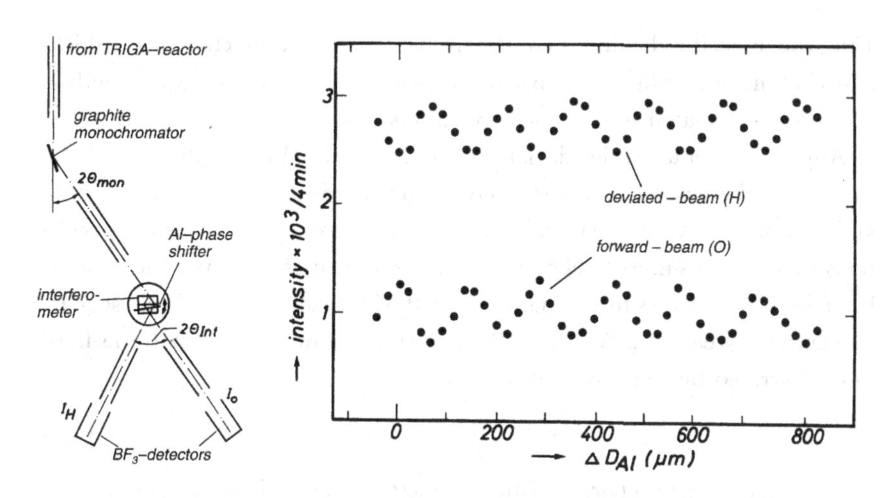

FIGURE 2.2 First observation of interference fringes with a perfect crystal interferometer [3].

A phase shift between the two coherent beams can be produced by nuclear, magnetic or gravitational interactions. In the first case, the phase shift is most easily calculated using the index of refraction [11, 12]:

$$n = \frac{k_{\text{in}}}{k_0} = 1 - \frac{\lambda^2 N}{2\pi}\sqrt{b_c^2 - \left(\frac{\sigma_r}{2\lambda}\right)^2} + i\frac{\sigma_r N\lambda}{4\pi}, \tag{1.1}$$

which simplifies for weakly absorbing materials ($\sigma_r \to 0$) to

$$n = 1 - \lambda^2 \frac{Nb_c}{2\pi}, \tag{1.2}$$

where b_c is the coherent scattering length and N is the particle density of the phase shifting material. As in ordinary light optics the change of the wave function is obtained as follows:

$$\psi \to \psi_0\, e^{i(n-1)kD} = \psi_0\, e^{-iNb_c\lambda D} = \psi_0\, e^{i\chi} = \psi_0\, e^{i\Delta\bar{k}}. \tag{1.3}$$

Therefore, the intensity behind the interferometer becomes

$$I_0 \propto |\psi_0^{\text{I}} + \psi_0^{\text{II}}|^2 \propto (1 + \cos\chi). \tag{1.4}$$

The intensity of the beam in the deviated direction follows from particle conservation:

$$I_0 + I_H = \text{const.} \tag{1.5}$$

Thus, the intensities behind the interferometer vary as a function of the thickness D of the phase shifter, the particle density N or the neutron wavelength λ. $\bar{\Delta}$ denotes the spatial shift of the wave trains.

Any experimental device deviates from the idealized assumptions made by the theory: the perfect crystal can have slight deviations from its perfectness, and its dimensions may vary slightly; the phase shifter contributes to imperfections by variations in its thickness and inhomogeneities; and even the neutron beam itself contributes to a deviation from the idealized situation because of its momentum spreads δk_i. Therefore, the experimental interference patterns have to be described by a generalized relation

$$I \propto A + B\cos(\chi + \phi_0), \tag{1.6}$$

where A, B and ϕ_0 are characteristic parameters of a certain set-up. It should be mentioned, however, that the idealized behaviour described by Eq. (1.4) can nearly be approached by a well balanced set-up [13]. The reduction of the contrast at high order results from the finite coherence lengths which are determined by the momentum spreads δk_i of the neutron beam. This causes a change in the amplitude factor of Eq. (1.6) as $B \to B\exp[-(\Delta_i \delta k_i)^2/2]$. Its characteristic lengths Δ_i^c determine the coherence length of the beam in direction i ($\Delta_i^c = 1/2\delta k_i$). The wavelength dependence of χ in Eq. (1.3) disappears in a special sample position where the surface of the sample is oriented parallel to the reflecting planes, and the path length through the interferometer becomes $D_0/\sin\Theta_B$ and, therefore, the phase $\chi = -2d_{hkl}Nb_c D_0$ becomes independent of the wavelength. In this case the damping at high interference orders due to the wavelength spread does not appear as in the standard position. Related results of a recent experiment where the interference pattern in the 256 interference order have been measured in the dispersive and the nondispersive sample position are shown in Fig. 2.3 [14]. The much higher visibility of the interferences in the nondispersive sample arrangement is visible and is caused by the much smaller momentum spread perpendicular to the reflecting planes caused by dynamical diffraction effects from the perfect crystal slabs.

All the results of interferometric measurements obtained up until now can be explained well in terms of the wave picture of quantum mechanics and the complementarity principle of standard quantum mechanics. Nevertheless, one should bear in mind that the neutron also carries well defined particle properties, which have to be transferred through the interferometer. These properties

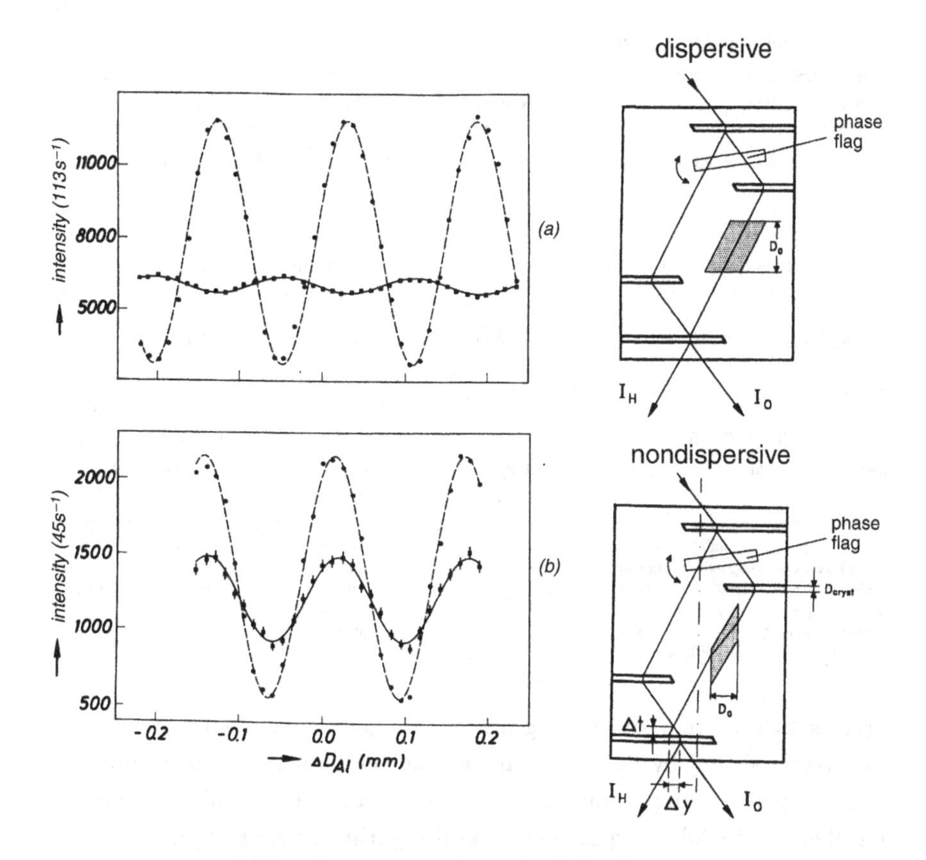

FIGURE 2.3 Interference pattern observed at high order ($m = 256$) with a dispersively (a) and a nondispersively (b) arranged sample [14] (dashed lines correspond to measurements at low order).

are summarized in Table 2.1 together with a formulation in the wave picture. Both particle and wave properties are well established and, therefore, neutrons seem to be a proper tool for testing quantum mechanics with massive particles, where the wave–particle dualism becomes obvious.

All neutron interferometric experiments pertain to the case of self-interference, where, during a certain time interval, only one neutron at most is inside the interferometer. Usually, at that time the next neutron has not yet been born and is still contained in the uranium nuclei of the reactor fuel. Although there is no interaction between different neutrons, they have a certain common history within predetermined limits which are defined, e.g., by the neutron moderation

Table 2.1. *Properties of the neutrons*

Particle properties	Connection	Wave properties
$m = 1.674928(1) \times 10^{-27}$ kg		$\lambda_c = \dfrac{h}{m.c} = 1.319695(20) \times 10^{-15}$ m
$s = \frac{1}{2}\hbar$	de Broglie	
$\mu = -9.6491783(18) \times 10^{-27}$ J/T	$\lambda_B = \dfrac{h}{m.v}$	for thermal neutrons $= 1.8$Å, 2200 m/s
$\tau = 887(2)$ s	Schrödinger	$\lambda_B = \dfrac{h}{m.v} = 1.8 \times 10^{-10}$ m
$R = 0.7$ fm	$H\psi(r,t) = i\hbar \dfrac{\delta\psi(\vec{r},t)}{\delta t}$	$\Delta_c = \dfrac{1}{2\delta k} \cong 10^{-8}$ m
$\alpha = 12.0(2.5) \times 10^{-4}$ fm^3 u-d-d-quark structure	& boundary conditions	$\Delta_p = v \cdot \Delta t \cong 10^{-2}$ m $\Delta_d = v \cdot \tau = 1.942(5) \times 10^6$ m $0 \leq \chi \leq 2\pi(4\pi)$
m: mass, s: spin, μ: magnetic moment, τ: β-decay lifetime, R: (magnetic) confinement radius, α: electric polarizability; all other measured quantities like electric charge, magnetic monopole and electric dipole moment are compatible with zero.	$-\mu B$ / μB ↕ two level system	λ_c: Compton wavelength, λ_B: de Broglie wavelength, Δ_c: coherence length, Δ_p: packet length, Δ_d: decay, δk: momentum width, Δt: chopper opening time, v: group velocity, χ: phase.

process, by their movement along the neutron guide tubes, by the monochromator crystal and by the special interferometer set-up. Therefore, any real interferometer pattern contains single particle and ensemble properties together. In the following sections typical experiments performed mainly by our group within the last 20 years will be presented.

2. Stochastic versus deterministic absorption

A certain beam attenuation can be achieved either by a semi-transparent material, by a proper chopper system, or by narrowing the beam aperture. The transmission probability in the first case is defined by the absorption cross section σ_a of the material $[a = I/I_0 = \exp(-\sigma_a ND)]$ and the change of the wave function is obtained directly from the complex index of refraction (Eq. 1.1):

$$\psi \to \psi_0\, e^{i(n-1)kD} = \psi_0\, e^{i\chi e^{-\sigma_a ND/2}} = e^{i\chi\sqrt{a}\psi_0}. \tag{2.1}$$

Therefore, the beam modulation behind the interferometer is obtained in the following form:

$$I_0 \propto |\psi_0^{\mathrm{I}} + \psi_0^{\mathrm{II}}|^2 \propto [(1+a) + 2\sqrt{a}\cos\chi]. \tag{2.2}$$

On the other hand, the transmission probability of a chopper wheel or another shutter system is given by the open to closed ratio, $a = t_{\text{open}}/(t_{\text{open}} + t_{\text{closed}})$, and one obtains after straightforward calculations

$$I \propto [(1-a)|\psi_0^{\mathrm{II}}|^2 + a|\psi_0^{\mathrm{I}} + \psi_0^{\mathrm{II}}|^2] \propto [(1+a) + 2a\cos\chi], \tag{2.3}$$

i.e. the contrast of the interference pattern is proportional to $\sqrt{a}$, in the first case, and proportional to a in the second case, although the same number of neutrons have been observed in both cases. The absorption represents a measuring process in both cases because a compound nucleus is produced with an excitation energy of several MeV, which is usually deexcited by captured gamma rays. These can easily be detected by different means.

Figure 2.4 shows a typical result for the transmission probabilities near to $a = 0.25$ as well as the dependence of the normalized contrast on the transmission probability [15, 16]. The different contrast becomes especially obvious for low transmission probabilities where the interfering part of the interference pattern is distinctly larger than the transmission probability through the semi-transparent absorber sheet. The discrepancy diverges for $a \to 0$ but it has been shown that in this regime the variations of the transmission due to variations of the thickness or of the density of the absorber plate have to be taken into account, which shifts the points below the $\sqrt{a}$-curve [18]. This can most easily be understood if the variation of the beam attenuation due to variations of the thickness or density fluctuations is included $a = \bar{a} + \Delta a$, which yields after averaging

$$\overline{\sqrt{a}} < \sqrt{\bar{a}} \tag{2.4}$$

indicating that the points fall below the $\sqrt{\bar{a}}$-curve.

The region between the linear and the square root behaviour can be reached by very narrow chopper slits or by narrow transmission lattices, where one starts to lose information of through which individual slit the neutron went. This is exactly the region which shows the transition between a deterministic and a stochastic view and, therefore, it can be formulated by a Bell-like [19, 20] inequality

$$\sqrt{a} > x > a. \tag{2.5}$$

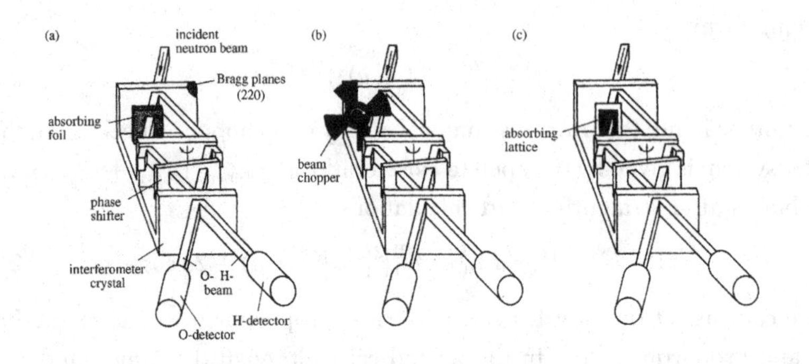
(a)
incident neutron beam
Bragg planes (220)
absorbing foil
phase shifter
interferometer crystal
O- H- beam
H-detector
O-detector
(b)
beam chopper
(c)
absorbing lattice

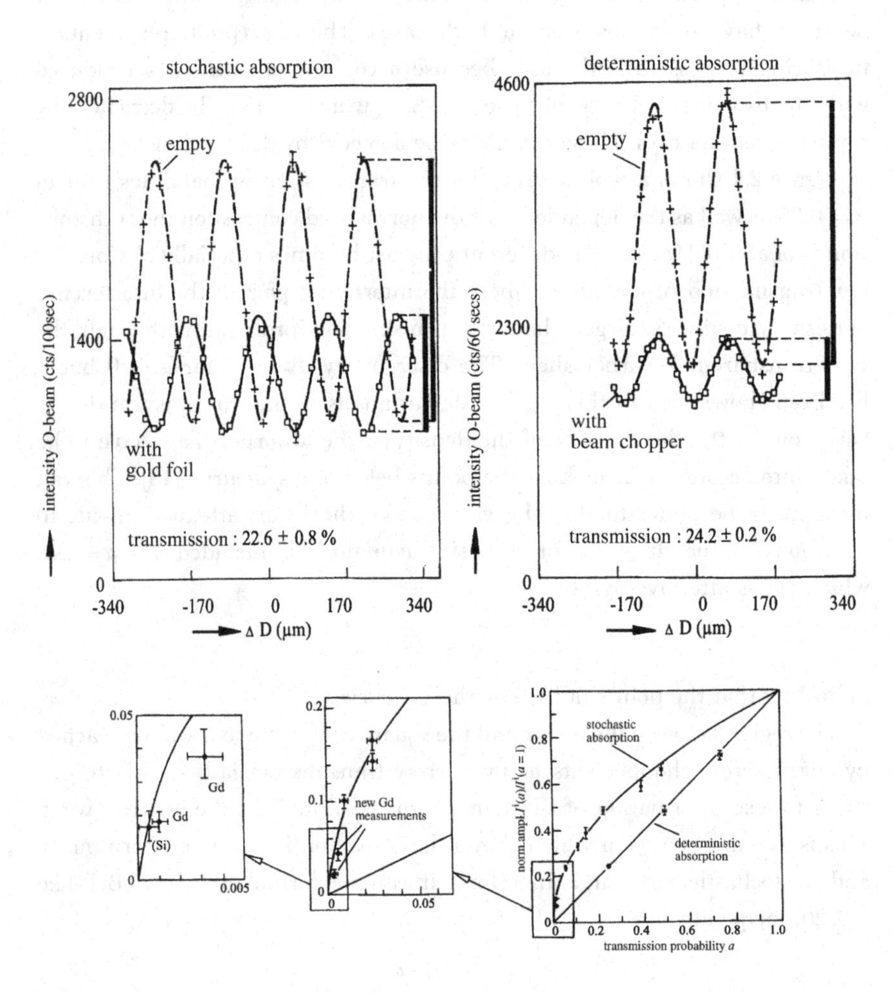
stochastic absorption
deterministic absorption
empty
with gold foil
with beam chopper
intensity O-beam (cts/100sec)
intensity O-beam (cts/60 secs)
transmission : 22.6 ± 0.8 %
transmission : 24.2 ± 0.2 %
Δ D (μm)
Gd
(Si)
new Gd measurements
stochastic absorption
deterministic absorption
norm.ampl.$I^{i}(a)/I^{i}(a=1)$
transmission probability a

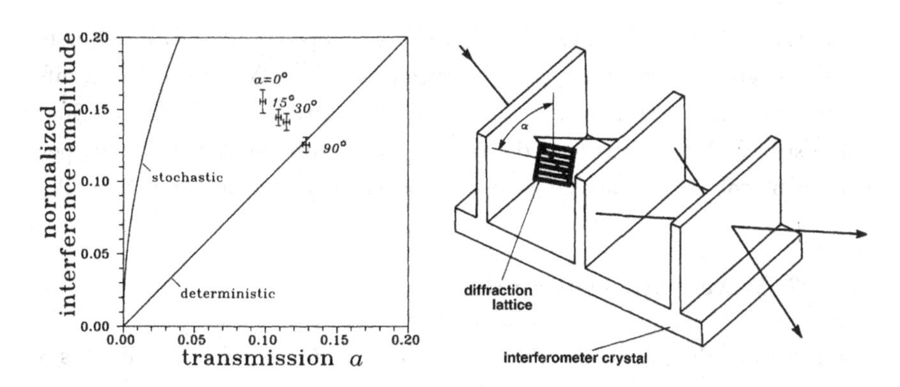

FIGURE 2.5 Lattice absorber in the interferometer approaching the quantum limit [16, 21].

The stochastic limit corresponds to the quantum limit when one does not know any more through which individual slit the neutron went. Which situation exists depends on how the slit widths l compares to the coherence lengths $(\Delta_i \sim (2\Delta k_i)^{-1})$ in the related direction. In the case that the slit widths become smaller than the coherence lengths, the wave function behind the slits shows distinct diffraction peaks which correspond to new quantum states $(n \neq 0)$, which now do not overlap with the undisturbed reference beam. The creation of the new quantum state means that those labelled neutrons carry information about the beam path chosen and, therefore, do not contribute to the interference amplitude [21]. A related experiment has been carried out by rotating an absorption lattice around the beam axis where one changes from $l \ll \Delta_x$ (vertical slits) to $l \gg \Delta_z$ (horizontal slits), Fig. 2.5, because the coherence length parallel to the reflecting lattice vector is much larger than in any other directions. Thus, the attenuation factor a has to be generalized, including not only nuclear absorption and scattering processes but also lattice diffraction effects if they remove neutrons from the original phase space.

FIGURE 2.4 (opposite) Sketch of the experimental arrangement for absorber measurements (above) (a) stochastic absorption, (b) deterministic absorption, (c) attenuation by a transmission grating. Typical results for stochastic and deterministic absorption (middle). Reduction of the contrast as a function of beam attenuation for different absorption methods (below) [16, 17].

A very similar situation exists if a very fast chopper produces beam bursts (packet lengths) shorter than the coherence time $\Delta t_c = \Delta/v$. In this case, diffraction in time occurs which also removes neutrons out of the original phase space. This limit is very difficult to reach with a mechanical chopper but it can probably be tackled with a high frequency spin flipper (section 6).

3. Wave–particle measure

The results of the previous section show that the degree of the contrast is a sensitive measure for the particle and wave character of the physical system. Therefore, an attempt is made to get a more general formulation, which accounts for stochastic and deterministic processes and for situations where a preparatory stage is created or where a real measuring process with a collapse of the wave field occurs.

For this more general discussion, the formulas can be used in the idealized form

$$I = T[1 + V \cos \phi'], \tag{3.1}$$

where we use the transparency of the system $T = (a+1)/2$ and the visibility of the interference fringes $V = V_s = 2a/(a+1)$ for the statistical case and $V = V_d = 2a^2/(a+1)$ for the deterministic case. Both quantities can be obtained from the interference pattern or by separate measurements of the beam attenuation factor a. The quantity $T.V$ denotes the amplitudes of the interference fringes as they are shown in Figs. 2.3 and 2.4. If the reduction of the contrast is also caused by high-order coherence phenomena or due to the roughness of the sample, the visibility factor in the equations becomes a product of the different visibility factors.

A measure for the particle nature can be found, if a quantity is taken which is the sum of the non-interfering intensity plus the probability of the neutrons being absorbed in beam I or II, respectively [22, 23]:

$$P(T.V) = T - T.V + (1 - T) = 1 - T.V = 1 - \langle |\psi^{\mathrm{I}}\psi^{\mathrm{II}}| \rangle, \tag{3.2}$$

and the wave nature from the amplitude of the interference pattern as

$$W(T.V) = T.V = \langle |\psi^{\mathrm{I}}\psi^{\mathrm{II}}| \rangle. \tag{3.3}$$

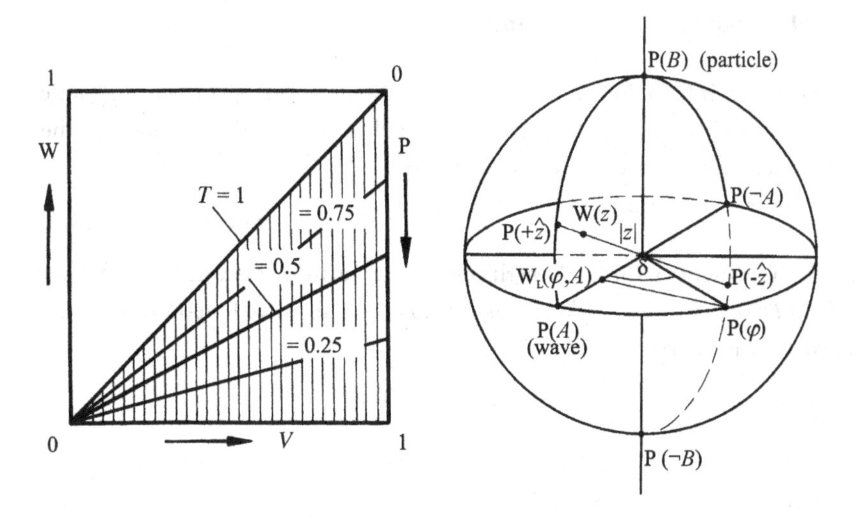

FIGURE 2.6 Synopsis of the wave–particle character in different interference experiments [22].

This rather simple formulation fulfils the relation

$$P + W = 1 \tag{3.4}$$

and can also be used for cases where the contrast attenuation is caused by large phase shifts which are of the order of the coherence length. In this case, the visibility varies but T remains 1. Obviously, this formulation is in complete agreement with the experimental results concerning the beam attenuation and the loss of contrast measurements discussed before. The whole result can be summarized in a single figure (Fig. 2.6). Measurements with equal beam attenuation in both beam paths belong to the line $V = 1$, high order loss of contrast and other deterministic phase mixing measurements are described by the curve $T = 1$ and many incoherent phase mixing processes lie near to the origin. This situation can be projected onto a sphere where the poles correspond to a pure particle-like and the equator corresponds to a pure wave-like behaviour [24, 25]. Decoherenced states lie inside the sphere. This kind of formulation may facilitate a combination with a Shannon information theoretic entropy approach, where various attempts have been made in the past.

4. 4π-symmetry of spinors

The magnetic interaction is caused by the dipole coupling of the magnetic moment of the neutron $\vec{\mu}$ to a magnetic field $\vec{B}$ ($H = -\vec{\mu}\vec{B}$). Therefore, the propagation of the wavefunction is given by

$$\psi \to \psi_0 \, \mathrm{e}^{-\mathrm{i}(Ht/\hbar)} = \psi_0 \, \mathrm{e}^{-\mathrm{i}(\mu Bt/\hbar)} = \psi_0 \, \mathrm{e}^{-\mathrm{i}\vec{\sigma}\vec{\alpha}/2} = \psi(\alpha), \tag{4.1}$$

where $\vec{\alpha}$ represents a formal description of the Larmor rotation angle around the field $\vec{B}$ ($\alpha = (2\mu/\hbar) \int B \, \mathrm{d}t \cong (2\mu/\hbar v) \int B \, \mathrm{d}s$). This wave function shows the typical 4π-symmetry of a spinor:

$$\begin{aligned} \psi(2\pi) &= -\psi(0) \\ \psi(4\pi) &= \psi(0), \end{aligned} \tag{4.2}$$

whereas 2π-symmetry exists only for the expectation values

$$|\psi(2\pi)|^2 = |\psi(0)|^2. \tag{4.3}$$

The 4π-periodicity becomes visible in interferometer experiments, as predicted theoretically [26–28], and as it has been verified experimentally in early neutron interferometric experiments [29, 30], where the intensity for unpolarized incident neutrons was found to be

$$I_0 \propto |\psi(0) + \psi(\alpha)|^2 \propto \left(1 + \cos\frac{\alpha}{2}\right). \tag{4.4}$$

The results of the first related experiment are shown in Fig. 2.7. These results are widely debated in the literature. It should be mentioned that this 4π-symmetry can always be attributed to real rotations in the case of fermions [31, 32]. Today, the most precise value for the periodicity factor is $\alpha_0 = 715.87 \pm 3.8$ degrees [33]. This value provides only a small margin for speculation about *SU*(2)-symmetry breaking, but a new and more precise determination of α_0 is recommended. The 4π-periodicity effect has been observed for unpolarized as well as polarized neutrons, which demonstrates the intrinsic feature of this phenomenon and the self-interference properties involved in this kind of experiment. New attention has been drawn to an interferometric observation of the Berry phase which represents a topological phenomenon and is, therefore, of central interest [34, 35]. In this case, the phase shift also depends around which axis the spin has been rotated or which interferometer loops were accessible for the neutrons. Recent interferometric experiments

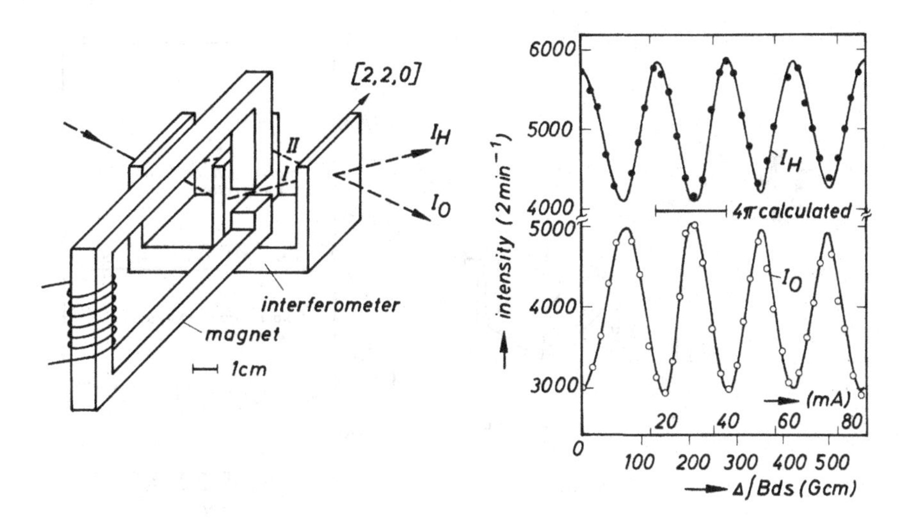

FIGURE 2.7 First observation of the 4π-symmetry factor of spinors [29].

verified these predictions and demonstrated again an additional amount of information stored in the wavefunction [36, 37].

5. Spin state interferometry

In this case, polarized neutrons are used, the polarization vector can be influenced differently in the two coherent beam paths and these beams can be superposed at the end of the interferometer. The principles of these experiments and the most important results are summarized in Fig. 2.8 [38, 39]. There is a marked difference between the action of a static flipper and a resonance flipper, which has to be discussed in more detail.

In the first case (static flipper) the wave function is changed by the flipper according to Eq. (3.1), which has to be applied for polarized incident neutrons:

$$\psi \rightarrow \mathrm{e}^{\mathrm{i}\chi}\,\mathrm{e}^{-\mathrm{i}\vec{\sigma}\vec{\alpha}/2}|z\rangle = \mathrm{e}^{\mathrm{i}\chi}\,\mathrm{e}^{-\sigma_y\pi/2}|z\rangle = -\mathrm{i}\sigma_y\,\mathrm{e}^{\mathrm{i}\chi}|z\rangle = \mathrm{e}^{\mathrm{i}\chi}|-z\rangle. \tag{5.1}$$

The rotation of the polarization vector around the y-axis has been postulated to be π [40]. Thus, two wave functions with opposite spin directions are superposed at the third plate

$$\psi \propto \left(|z\rangle + \mathrm{e}^{\mathrm{i}\chi}|-z\rangle\right), \tag{5.2}$$

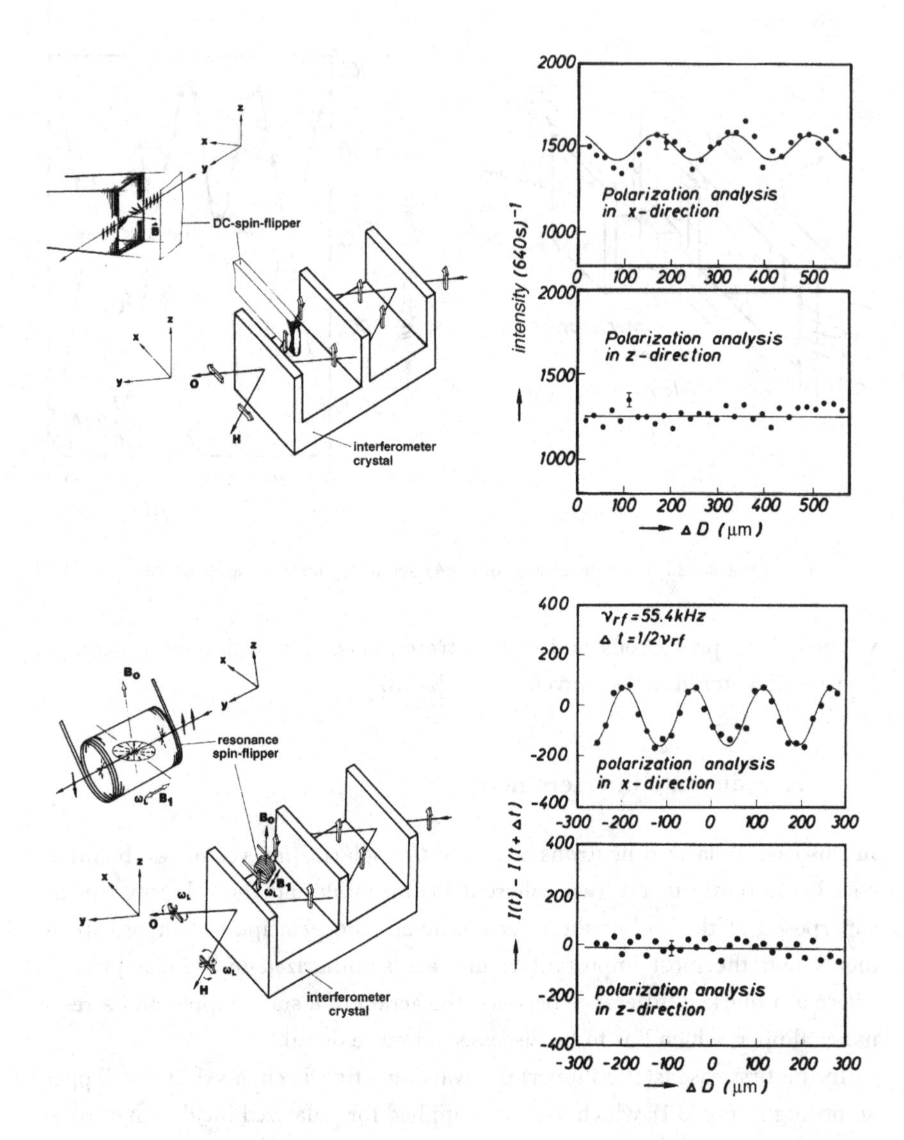

FIGURE 2.8 Sketch of the static (above) and the time-dependent (below) spin superposition experiment with characteristic results [38, 39].

which corresponds to the situation proposed by Wigner [41] in 1963 to verify the quantum mechanical spin superposition law. In this case the intensities in the O- and H-beams are equal, and the beams are polarized in the (x, y)-plane, i.e. perpendicular to both the initial spin directions. The angle of the polarization in the (x, y)-plane is given by the nuclear phase shift,

$$\vec{P}_0 = \frac{\langle\psi|\vec{\sigma}|\psi\rangle}{I_0} = \begin{pmatrix} \cos\chi \\ \sin\chi \\ 0 \end{pmatrix}. \tag{5.3}$$

Thus, a pure initial state is transferred to a pure final state which is different from both the states existing before superposition. The interference pattern appears only if a polarization analysis is performed in the $|x\rangle$- or in the $|y\rangle$-direction. If the analyser is set in the $|z\rangle$- (or $|-z\rangle$)-direction, no intensity modulation is observed.

The second version of the spin superposition experiment was performed with a Rabi-type resonance flipper which is also commonly used in polarized neutron physics. This kind of interaction is time-dependent and, in addition to the spin-inversion, an exchange of the resonance energy $E_{\mathrm{HF}} = \hbar\omega_{\mathrm{r}}$ occurs between the neutron and the resonator system, which has to be considered in the interferometric experiment. This energy exchange was observed in a separate experiment, where the energy resolution ΔE of the apparatus was better than the Zeeman energy splitting $(\Delta E < E_{\mathrm{HF}})$ [42]. This experiment was performed according to a proposal of Drabkin and Zhitnikov [43]. For a complete spin reversal the frequency of the field has to match the resonance condition, and the amplitude B_1 has to fulfil the relation $|\mu|B_1 l/\hbar v = \pi$, where l is the length of the coil. Oscillating fields are used instead of purely rotational fields and, therefore, only one component contributes to the resonance, which causes a slight shift of the resonance frequency from the Larmor frequency $\omega_{\mathrm{L}} = 2|\mu|B_0/\hbar$ due to the Bloch–Siegert effect $(\omega_{\mathrm{r}} = \omega_{\mathrm{L}}[1 + (B_1^2/16B_0^2)])$ [44, 45]. Thus, the wave function of the beam with the flipper changes according to

$$\psi \rightarrow \mathrm{e}^{\mathrm{i}\chi}\,\mathrm{e}^{(\omega-\omega_{\mathrm{r}})t}|-z\rangle. \tag{5.4}$$

Therefore, a spin-up and a spin-down state are superposed at the position of the third plate. The final polarization of the beam in the forward direction is given by

$$\vec{P} = \begin{pmatrix} \cos(\chi + \omega_{\mathrm{r}} t) \\ \sin(\chi + \omega_{\mathrm{r}} t) \\ 0 \end{pmatrix} \tag{5.5}$$

and lies again in the (x, y)-plane, but now rotates within this plane with the resonance (Larmor) frequency without being driven by a magnetic field. A stroboscopic method was needed for the observation of this effect. The direction of the polarization in the (x, y)-plane depends on the status (phase) of the resonance field and, therefore, has to be measured synchronously with this phase.

The observed interference pattern (Fig. 2.8) demonstrates that coherence persists, although a well defined energy exchange between the neutron and the apparatus exists. Thus, an energy exchange is not automatically a measuring process. As we will see later on, the exchanged photon cannot be used for a quantum nondemolition measurement. In our experiment, the following argument based on different uncertainty relations can be used: firstly, one single absorbed or emitted photon of the resonator cannot be detected because of the photon number–phase uncertainty relation, which can be written in the form [46, 47]

$$(\Delta N)^2 \frac{(\Delta S)^2 + (\Delta C)^2}{\langle S\rangle^2 + \langle C\rangle^2} \geq 1/4, \tag{5.6}$$

where S and C can be expressed by the creation and annihilation operators, $C = (a_- + a_+)/2$ and $S = (a_- - a_+)/2\mathrm{i}$, whose matrix elements couple coherent Glauber states. For our purpose this relation can be used in its simpler form

$$\langle \Delta N^2 \rangle \langle \Delta\theta^2 \rangle \geq 1/4. \tag{5.7}$$

The uncertainty of the photon number of the resonator is minimized for a coherent state resonator by $\Delta N \cong \sqrt{\langle N\rangle}$ [48] and, therefore, the lower limit for the phase uncertainty becomes $\Delta\theta \cong 1/(2\sqrt{\langle N\rangle})$. Because in this kind of spin-superposition experiment the phase determination of the flipper field is required to be better than $\Delta\theta < \eta$ for the stroboscopic method, it is impossible to observe a single absorbed or emitted photon $(\Delta N \geq 1)$.

A second version of the beam path detection may be based on the observation of the energy change of the neutron. This can only be achieved if the energy resolution of the instrument fulfils the relation $\Delta E \leq 2\mu B_0$. On the other hand, the stroboscopic measuring method requires time channels $\Delta t \leq 1/2\nu_{\mathrm{HF}} = h/4|\mu|B_0$, which provides another constraint on the experiment. Both conditions cannot be fulfilled with respect to the energy–time uncertainty relation concerning the beam parameters $\Delta E \Delta t \geq \hbar/2$. Therefore, we conclude that a simultaneous detection of the beam path through the interferometer and of the interference pattern remains impossible. This kind of experiment has also been analysed in terms of a coherent state or of a number state resonator [49]. These authors came to the same conclusion, that interference becomes destroyed if a path signal is extracted from the resonators.

It has been proposed by Vigier's group [50, 51] that new information about the particle–wave duality can be obtained with resonator coils in both coherent beams. The corresponding experiments will be discussed in the next section.

6. Double coil experiments

The experimental arrangement for the double coil experiment is shown in Fig. 2.9 [52, 53]. The final polarization lies in the $|-z\rangle$ direction, and the energy transfer $h\omega_r$ can be smaller or larger than the energy resolution ΔE because this information cannot be in any way associated with a beam path detection. The lay-out of the experiment followed the proposal of Vigier's group [50, 51]. According to our previous considerations, the change of the wave functions with the resonance flippers tuned to the resonance frequency can be written for polarized incident neutrons ($|z\rangle$) and for different modes of operation as follows.

(a) Both flippers are operated synchronously without a phase shift between the flipper fields:

$$\psi \rightarrow e^{i(\omega-\omega_r)t}|-z\rangle + e^{i\chi}\, e^{i(\omega-\omega_r)t}|-z\rangle. \tag{6.1}$$

This results in an intensity modulation

$$I_0 \propto 1 + \cos\chi, \tag{6.2}$$

which is independent of the flipper fields.

(b) Both flippers are operated synchronously with a distinct phase shift Δ:

$$\psi_0 \rightarrow e^{-i(\omega-\omega_r)t}|-z\rangle + e^{i\chi}\, e^{i\Delta}\, e^{i(\omega-\omega_r)t}|-z\rangle. \tag{6.3}$$

In this case, the intensity modulation is given by

$$I_0 \propto 1 + \cos(\chi + \Delta). \tag{6.4}$$

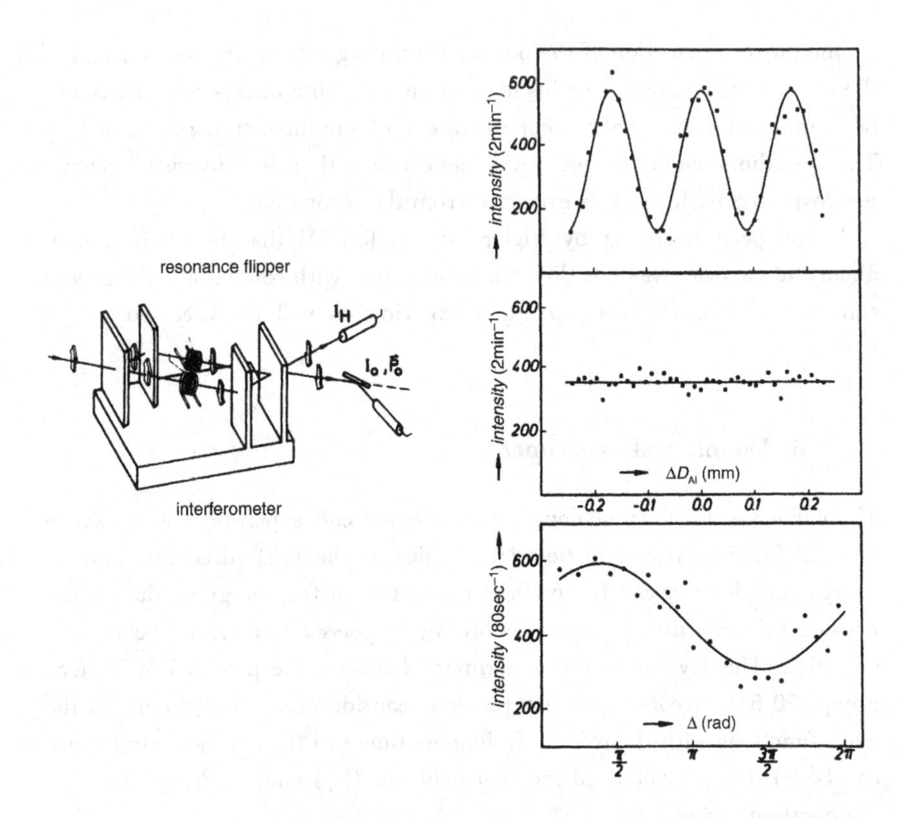

FIGURE 2.9 Sketch of the double resonance coil experiment (right) and results of the double coil experiments (left) [52, 53]. Top: Synchronous flipper fields with $\nu_r = 71.90\,\text{kHz}$ (Eq. (4.1)). Middle: Two slightly fluctuating independent flipper fields with $\nu_r = 71.92 \pm 0.02\,\text{kHz}$ (Eq. (4.3)). Bottom: Interference pattern as a function of the phase shift Δ between both flipper fields at $\nu_r = 71.90\,\text{kHz}$ (Eq. (4.4)).

(c) Both flippers are operated asynchronously with statistically fluctuating phase differences $\Delta(t)$ which average out during the measuring interval. Then

$$I_0 \propto \text{const.} \tag{6.5}$$

It should be mentioned in this context that, even in this case, coherence phenomena can be observed if a stroboscopic investigation is performed ($I_0 = I_0(\Delta)$).

The results of these related experiments are shown in Fig. 2.9. Complete agreement with the theoretical predictions is found. The interference

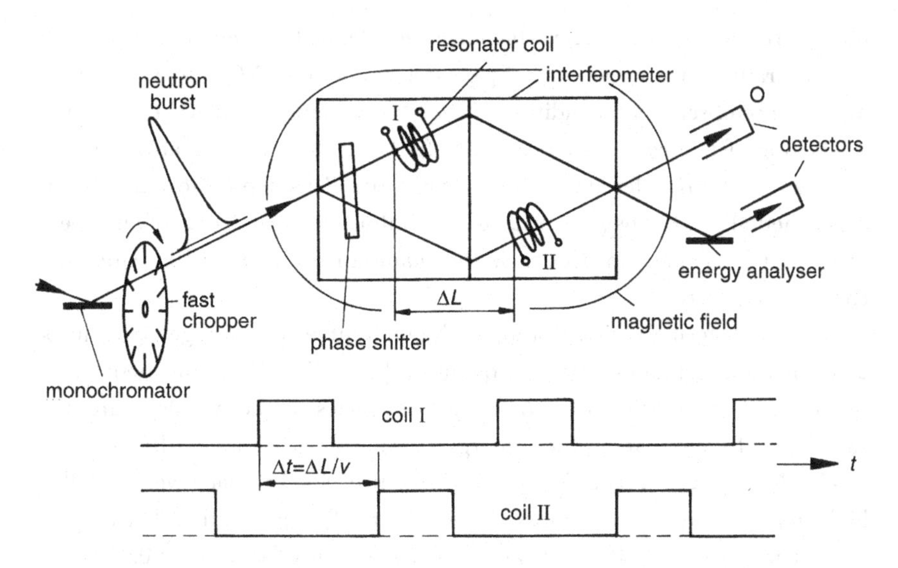

FIGURE 2.10 Proposed double coil experiment where pulsed neutron beams and pulsed neutron flippers are used and where interference and energy change of the neutrons can be observed simultaneously [54].

properties are preserved, although an energy exchange $\hbar\omega_r$ certainly takes place. Only quanta within a narrow energy band around $\hbar\omega_r$ and no others are excited inside the flipper resonator. Therefore, one could believe that the spin flip and the energy transfer process to the neutron occurred inside one of the two coils, which would demonstrate that the neutron has chosen one of the two possible paths. But even this rather weak statement would require the concept of pilot waves, quantum potentials, etc., leading immediately to questions about the interpretation of quantum mechanics, which are not the primary subject of this article. In this connection, an experiment is proposed which shows this situation even more clearly [54] (Fig. 2.10). In this case, the flippers are operated only if the neutron burst is inside the coils and a strong magnetic field causes an energy shift $\Delta E_{\mathrm{HF}} = 2\mu B_0$ which is smaller or larger than the energy width of the beam (ΔE). In contrast to the conclusion drawn in [54], it has to be stated that the interference pattern exists only in cases $\Delta E_{\mathrm{HF}} < \Delta E$, whereas in cases $\Delta E_{\mathrm{HF}} > \Delta E$, i.e. for very strong magnetic fields or a very high resolution, the interference pattern disappears due to a quantum chopping effect. In this case, the oscillating field varies faster than the coherence time $\Delta t_c = \Delta / v$ which has been defined in Section 3. This results in a contradiction between virtual path determination ($\Delta E_{HF} > \Delta E$) and the interference observation condition ($1/\nu_{HF} > \Delta t_c$), and means that as soon as a labelling of neutrons exists ($\Delta E_{HF} > \Delta E$) – which is a virtual beam path detection – the interference pattern disappears. This shows how closely the experimental conditions (ΔE, or Δ, or Δt_c) influence the outcome of an experiment and how one can approach the quantum limit where the situations change drastically.

In the interfering regime ($\Delta E_{HF} < \Delta E$) a double coil arrangement can be used for the observation of a new quantum beat effect. If the magnetic guide fields are slightly different (ΔB_0), the frequencies of the two coils are also slightly different and the energy transfer becomes different too ($\Delta E = \hbar(\omega_{r1} - \omega_{r2})$). The frequency difference can be made very small if high quality frequency generators are used for the field generation. The flipping efficiencies for both coils are always very close to unity (better than 0.99). Now, the wave functions change according to

$$\psi \rightarrow \mathrm{e}^{\mathrm{i}(\omega - \omega_{r1})t} |-z\rangle + \mathrm{e}^{\mathrm{i}\chi} \mathrm{e}^{\mathrm{i}(\omega - \omega_{r2})t} |-z\rangle. \tag{6.6}$$

Therefore, the intensity behind the interferometer exhibits a typical quantum beat effect given by

$$I \propto 1 + \cos[\chi + (\omega_{r1} - \omega_{r2})t]. \tag{6.7}$$

Thus, the intensity behind the interferometer oscillates between the forward and deviated beam without any apparent change inside the interferometer [52, 53]. The time constant of this modulation can reach a macroscopic scale which is correlated to an uncertainty relation $\Delta E \Delta t \leq \hbar/2$. Figure 2.11 shows the result of an experiment, where the periodicity of the intensity modulation, $T = 2\pi/(\omega_{r1} - \omega_{r2})$, amounts to $T = (47.90 \pm 0.15)\,\mathrm{s}$, caused by a frequency difference of about 0.02 Hz. This corresponds to a mean difference ΔE of the energy transfer between the two beams, $\Delta E = 8.6 \times 10^{-17}$ eV, and to an energy sensitivity of 2.7×10^{-19} eV, which is better by many orders of magnitude than that of other advanced spectroscopic methods. This high resolution is strongly decoupled from the monochromaticity of the neutron beam, which was $\Delta E_B \cong 5.5 \times 10^{-4}$ eV around a mean energy of the beam $E_B = 0.023\,\mathrm{eV}$ in this case. It should be mentioned that the result can also be interpreted as being the effect of a slowly varying phase $\Delta(t)$

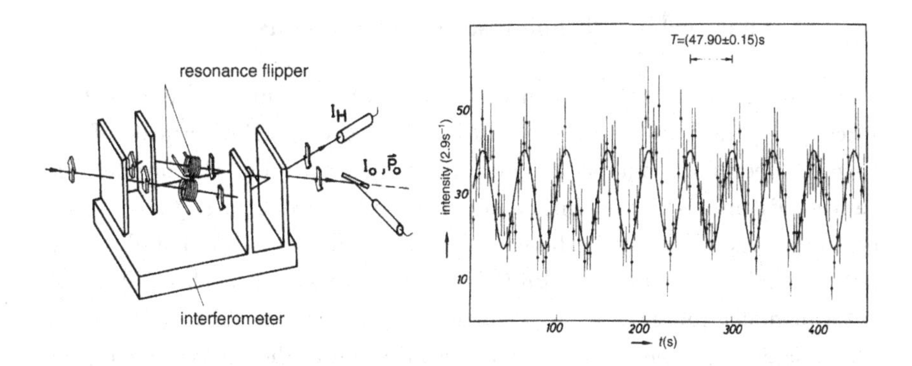

FIGURE 2.11 Quantum beat effect observed when the frequencies of the two flipper coils differ by about 0.02 Hz around 71.89979 kHz [53].

between the two flipper fields (see Eq. (5.4)), but the more physical description is based on the argument of a different energy transfer. The extremely high resolution may be used for fundamental, nuclear and solid state physics applications.

The quantum heat effect can also be interpreted as a magnetic Josephson effect analogue [55]. In this case, the phase difference is driven by the magnetic energy

$$\frac{\delta}{\delta t}(\Delta_2 - \Delta_1) = \omega_{r2} - \omega_{r1} = \frac{1}{\hbar}2\mu\Delta B_0, \tag{6.8}$$

which yields the observed modulation (compare with Eq. (6.7))

$$I \propto (1 + \cos\Delta(t)), \tag{6.9}$$

where $\delta(t) = 2\mu\Delta B_0 t/\hbar$. ΔB_0 denotes the difference of the guide fields in both beams. This is analogous to the well-known Josephson effect in superconducting tunnel junctions [56], where the phases of the Cooper pairs in the two superconductors is related according to

$$\frac{\delta}{\delta t}(\phi_2 - \phi_1) = \frac{1}{\hbar}(E_2 - E_1) = \frac{1}{\hbar}2\,\mathrm{eV}, \tag{6.10}$$

which is driven by the electrical potential V between the superconductors. More recently, multiphoton exchange has been observed between the neutron and an oscillating magnetic field detuned from resonance [57].

7. Phase echo and time-resolved experiments

Such systems are similar to spin echo systems known in advanced neutron spectroscopy [58], but use the phase of the wave function instead of the Larmor precession angle as the measurable quantity [59]. The interference pattern disappears if the shift of the wave packets due to a phase shifter becomes larger than the coherence length of the beam in that direction ($\Delta^c = 1/2\delta k = \lambda^2/4\pi\delta\lambda$; see also Fig. 2.3). This behaviour has been observed experimentally [60, 61]. By applying an opposite phase shift in the same beam, or the same phase shift in the second beam of the interferometer, the disappeared interference pattern can be recovered to full contrast, as shown schematically in Fig. 2.12. Such experiments can demonstrate that a phase information can exist although the measured signal looks like a statistical mixture. The coherence properties can be recovered if a proper measuring method is applied. The recovering of a smeared out interference pattern has been demonstrated recently by using combinations of thick Bi- and Ti-phase shifters [62]. This method establishes a new horizon to retrieve the interference properties of a beam whose wave packets have been shifted more than its coherence length. It shows that interference can be generally recovered, at least to a large extent, if we know how we have to apply different measuring methods.

Another method of recovering an interference pattern exists by increasing the resolution by time-of-flight methods. A schematic arrangement is shown in Fig. 2.13. The motion of a free wave packet is described by the time-dependent Schrödinger equation. In the case of minimum uncertainty packets with Gaussian widths δx and δk in real and momentum space ($\delta x \delta k = 1/2$) one

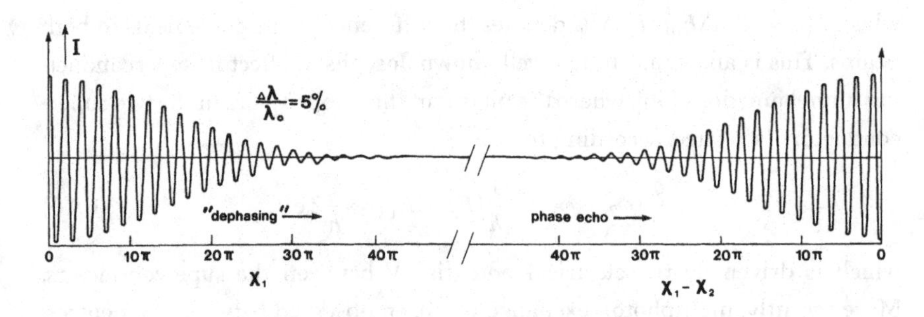

FIGURE 2.12 Principle of a phase echo system [59].

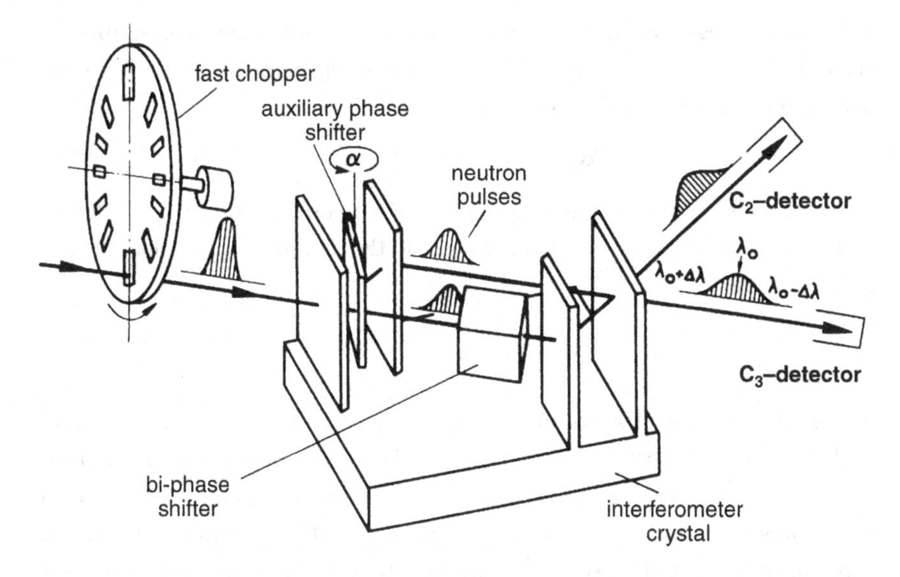

FIGURE 2.13 Sketch of the apparatus for interference experiments with pulsed beams.

expects the quantum mechanical spreading as a function of time (e.g. [63]):

$$[\delta x(t)]^2 = [\delta x(0)]^2 + \left[\frac{(\hbar/2m)t}{\delta x(0)}\right]^2. \tag{7.1}$$

This minimum uncertainty wave packet is difficult to achieve because $\delta x(0)$ has to approach the coherence length $\Delta_c \sim (2\delta k)^{-1}$, as discussed previously. That also means that pulse lengths of the order of the coherence time $\Delta t^c = \Delta^c/v$ have to be produced, which means chopper opening times of the order of ns. In this case, diffraction in time would play an important role, which is a complete analogue to the known single slit diffraction phenomena in real space [64, 65]. Thus, Fraunhofer- and Fresnel-like phenomena are expected to occur, which changes the energy of the beam accordingly. For a triangular slit opening for a time Δt one expects for an incident plane wave with a frequency ω_0 a frequency spectrum

$$|\phi(\omega)|^2 \propto \left[\frac{\sin(\omega - \omega_0)\Delta t}{\omega - \omega_0}\right]^2. \tag{7.2}$$

For long opening times ($\Delta r \gg \Delta t^c$) diffraction effects occur at the edges only, which can be neglected in most cases. Then the phenomena can be described by classical distribution functions, but their spreading happens similarly to the quantum case. For Gaussian pulses one gets

$$[\delta x(t)]^2 = [\delta x(0)]^2 + [\delta v.t]^2, \tag{7.3}$$

where one notices the similarity to Eq. (7.1) if one uses the de Broglie relation and $\Delta_c = (2\delta k)^{-1} = \lambda^2/4\pi\delta\lambda$. In terms of the temporal pulse length one obtains

$$[\Delta t]^2 = [\Delta t_p]^2 + \left[\frac{\delta\lambda}{\lambda} t_0\right]^2, \tag{7.4}$$

where Δt_p is the opening time of the chopper and $t_0 = L/v_0$ is the average time-of-flight of neutrons between the chopper and the detector at distance L. In fact, Eq. (7.4), which describes the broadening of the intensity pulse, is identical to the expression which describes the broadening of a quantum mechanical wave packet (Eq. (7.1)) except for the fact that in Eq. (7.4) $\delta x(0)$ does not fulfil the minimum uncertainty relation. The mean wavelength of neutrons being measured at time t at a detector with an effective thickness Δz becomes

$$\langle \lambda(t) \rangle = \lambda_0[1 + t_0(t - t_0)(\delta\lambda/\lambda_0)^2/(\Delta t')^2], \tag{7.5}$$

with

$$[\Delta t']^2 = [\Delta t]^2 + [\Delta z/v_0]^2.$$

A very similar formula holds for the mean phase shift $\langle \chi(t) \rangle$ which is proportional to $\langle \lambda(t) \rangle$. The linear flight-time dependence of $\langle \chi(t) \rangle$ has been observed in [62].

At a position L at a time t there still exists in a certain time channel a Gaussian distribution with a restricted spectral width given by

$$\left(\frac{\delta\lambda}{\lambda_0}\right)^{\prime 2} = \left[\left(\frac{\lambda_0}{\delta\lambda}\right)^2 + \left(\frac{t_0}{\Delta t_p'}\right)^2\right]^{-1}, \tag{7.6}$$

which causes less interference attenuation for measurements with narrow time channels than for measurements with the whole beam and permits the recovering of the smeared out interference pattern [62, 66] (Fig. 2.14(a) and (b)).

We now know how neutron pulses spread as they propagate. At larger distances downstream the edges of each pulse begin to overlap incoherently with the preceding and following one. When Gaussian shaped neutron pulses are

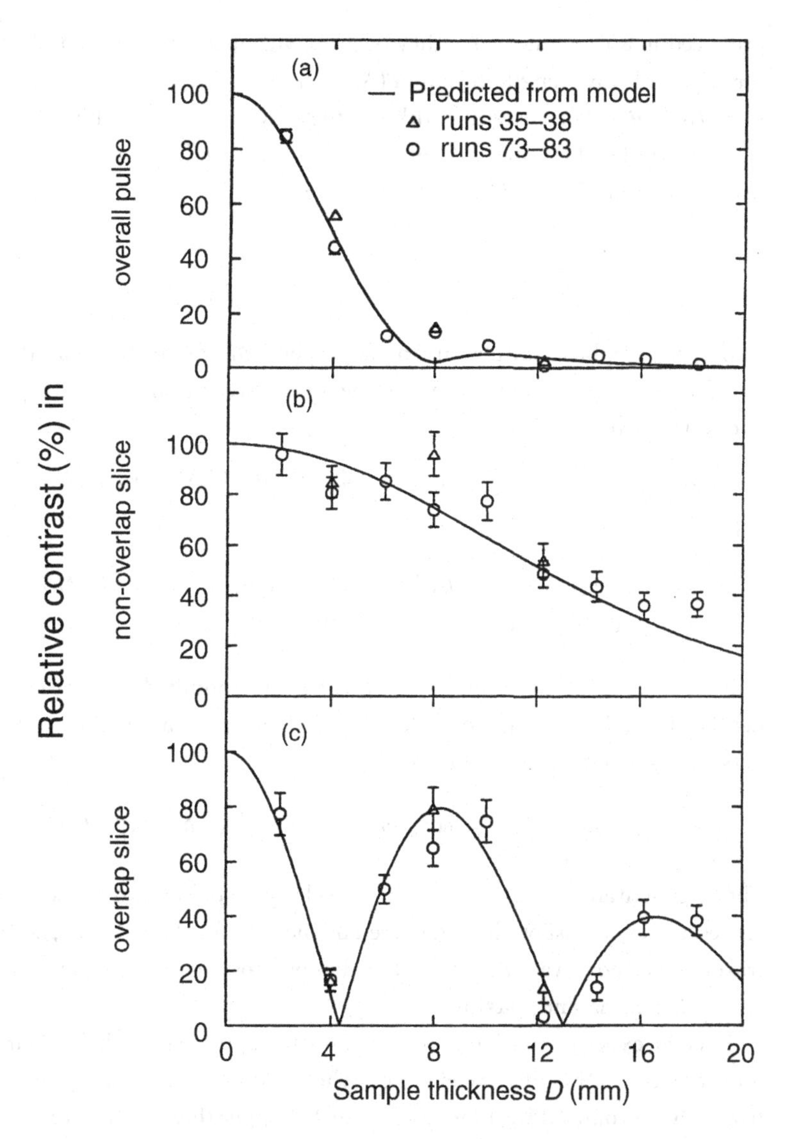

FIGURE 2.14 The relative contrasts measured in the overall pulse versus the sample thickness D and the predicted curve from a model based on a double Gaussian spectral fit to the actual wavelength distribution (a). The relative contrasts measured versus the sample thickness D in the non-overlap time slice (b), and for the overlap time slice (c) [66].

produced in a time interval T they overlap with neutron wave length equal to the standard wave length deviation ($\lambda_1 = \lambda_0 - \delta\lambda$, $\lambda_2 = \lambda_0 + \delta\lambda$) at a distance $L^c = Th/2m\delta\lambda$. The spectral width at the related time interval is given by the spectral widths at λ_1 and λ_2 (Eq. (7.6)). The measured intensity in the related time slide around $t = t_0^c(1 - \delta\lambda/\lambda_0)$ is given as

$$I(\lambda_1, \lambda_2) = I(\lambda_1) + I(\lambda_2)$$
$$\propto 2 + \langle|\Gamma(\Delta_1)|\rangle \cos\chi_1 + \langle|\Gamma(\Delta_2)|\rangle \cos\chi_2. \tag{7.7}$$

In the case of phase shifters arranged perpendicularly to the beam (Fig. 2.13) and for rather narrow Gaussian wavelength distributions around λ_1 and λ_2, one can reformulate Eq. (7.6):

$$I(\lambda_1, \lambda_2) \propto 1 + \exp[-\chi_0^2(\delta\lambda/\lambda_0)^2/2] \cos(Nb_cD\Delta\lambda) \cdot \cos(Nb_cD\bar{\lambda}) \tag{7.8}$$

with

$$\chi_0 = -Nb_cD\lambda_0 \qquad \langle|\Gamma(\Delta_1)|\rangle \approx \langle|\Gamma(\Delta_2)|\rangle = -\exp[-(\Delta\delta k)^2/2]$$
$$\bar{\lambda} = (\lambda_1 + \lambda_2)/2 \qquad \Delta\lambda = (\lambda_1 - \lambda_2)/2.$$

Thus, the total intensity exhibits a series of 'beats', modified by a decaying exponential. The relative contrast of the intensity pattern as a function of the thickness of the phase shifter reads as

$$C(D) = \frac{I_{\mathrm{Max}}}{I_{\mathrm{Min}}} = |\cos(Nb_cD\Delta\lambda)| \exp[-\chi_0^2(\delta\lambda/\lambda_0)^2/2]. \tag{7.9}$$

The exponential term is due to the wavelength spread and the cosine term represents a contrast oscillation caused by the overlap of two pulses, i.e. the contrast is expected to vary according to a damped cosine function going through a series of minima and maxima.

Related experiments have indeed shown this oscillatory behaviour of the interference contrast in time channels where pulse overlap exists [66]. The neutron pulses produced by a fast mechanical chopper (Fig. 2.13) were on the one hand significantly shorter than the dimensions of the perfect crystal interferometer, thus assuring that there is no permanent overlap of wave functions or distribution functions at the beam splitting and beam overlapping part of the interference experiment. The experiment again demonstrates very clearly the single particle interference phenomena in neutron interference experiments, because the mean occupation number in a single neutron pulse is much smaller than unity, i.e. of the order of 10^{-4}. The neutron pulses used in that experiment are, on the other hand, considerably longer than the coherence length of the beams and, therefore, diffraction in time effects are negligible. Several aspects of neutron wave mechanics have been explored and have been shown to be in agreement with the theoretical predictions. The spreading of the neutron pulses has been observed as well as the spectral narrowing in distinct time slices downstream of the beam.

8. Postselection of momentum states

In the course of several neutron interferometer experiments [66–69] it has been established that smoothed out interference properties at high interference order can be restored, even behind the interferometer, when a proper spectral filtering is applied. The experimental arrangement with an indication of the wave packets at different parts of the interference experiment is shown in Fig. 2.15. An additional monochromatization is applied behind the interferometer by means of various single crystals brought into Bragg position or by time-of-flight systems. Using wave packets for the wave functions, the

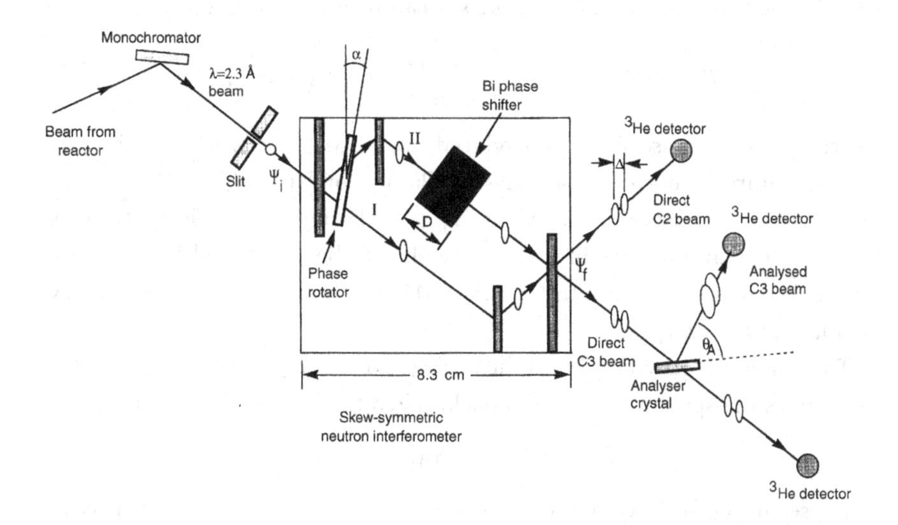

FIGURE 2.15 Experimental arrangement for postselection of momentum states.

momentum-dependent intensity reads as:

$$I_0(\vec{r}, \vec{k}) = |\psi_0^{\mathrm{I}}(\vec{r}, \vec{k}) + \psi_0^{\mathrm{II}}(\vec{r} + \Delta\vec{k})|^2 \propto g(\vec{k})(1 + \cos(\vec{\Delta}(\vec{k}) \cdot \vec{k})), \quad (8.1)$$

whereas the overall beam reads as

$$I_0(\vec{\Delta}_0) \propto 1 + |\Gamma(\vec{\Delta}_0)| \cos \vec{\Delta}_0 \cdot \vec{k}_0, \quad (8.2)$$

where $\vec{\Delta}_0$ represents the spatial phase shift for the $\vec{k}_0$-component of the packet. Equation (8.2) describes the interference fringes when $\vec{\Delta}_0$ is varied. $|\Gamma(\vec{\Delta})|$ denotes the coherence function which reads for Gaussian packets $|\Gamma(\vec{\Delta})| \propto \exp[-(\Delta_i \delta k_i)^2/2]$ (compare comment after Eq. (1.6)). Its definition is

$$\Gamma(\vec{\Delta}) = \langle \psi^*(0)\psi(\vec{\Delta}) \rangle, \quad (8.3)$$

i.e. it is the auto-correlation function of the wave function. The formula also shows that the overall interference fringes disappear for spatial phase shifts much larger than the coherence lengths $[\Delta_i \geq \Delta_i^c = 1/(2\delta k_i)]$. This behaviour is shown in Fig. 2.16 and has been verified experimentally by several investigations for Gaussian and non-Gaussian neutron beams [60, 61].

In our experiment we deal with the coherence properties along the interferometer axis (x), where the (tangential) components of the momentum vectors (and coherence lengths) do not change due to Bragg diffraction. According to basic quantum mechanical laws, the related momentum distribution follows from Eq. (8.2) and for Gaussian packets it can be rewritten in the form

$$I_0(k) = \exp[-(k - k_0)^2/2\delta k^2]\left\{1 + \cos\left(\chi_0 \frac{k_0}{k}\right)\right\}, \quad (8.4)$$

where the mean phase shift is introduced ($\chi_0 = k_0\Delta_0 = Nb_c\lambda_0 D_{\mathrm{eff}}$). The surprising feature is that $I_0(k)$ becomes oscillatory for large phase shifts where the interference fringes described by Eq. (8.2) disappear (see Fig. 2.16). This indicates that interference in phase space has to be considered [71, 72]. The second beam behind the interferometer (H) just shows the complementary modulation $I_{\mathrm{H}} = I_{\mathrm{total}} - I_0$.

The amplitude function [73] of the packets arising from beam paths I and II determines the spatial shape of the packets behind the interferometer,

$$I_0(x) = |\psi(x) + \psi(x + \Delta)|^2, \quad (8.5)$$

which separates for large phase shifts into two peaks (Fig. 2.16). For Gaussian packets, having a spatial width, δx, which corresponds to the coherence length

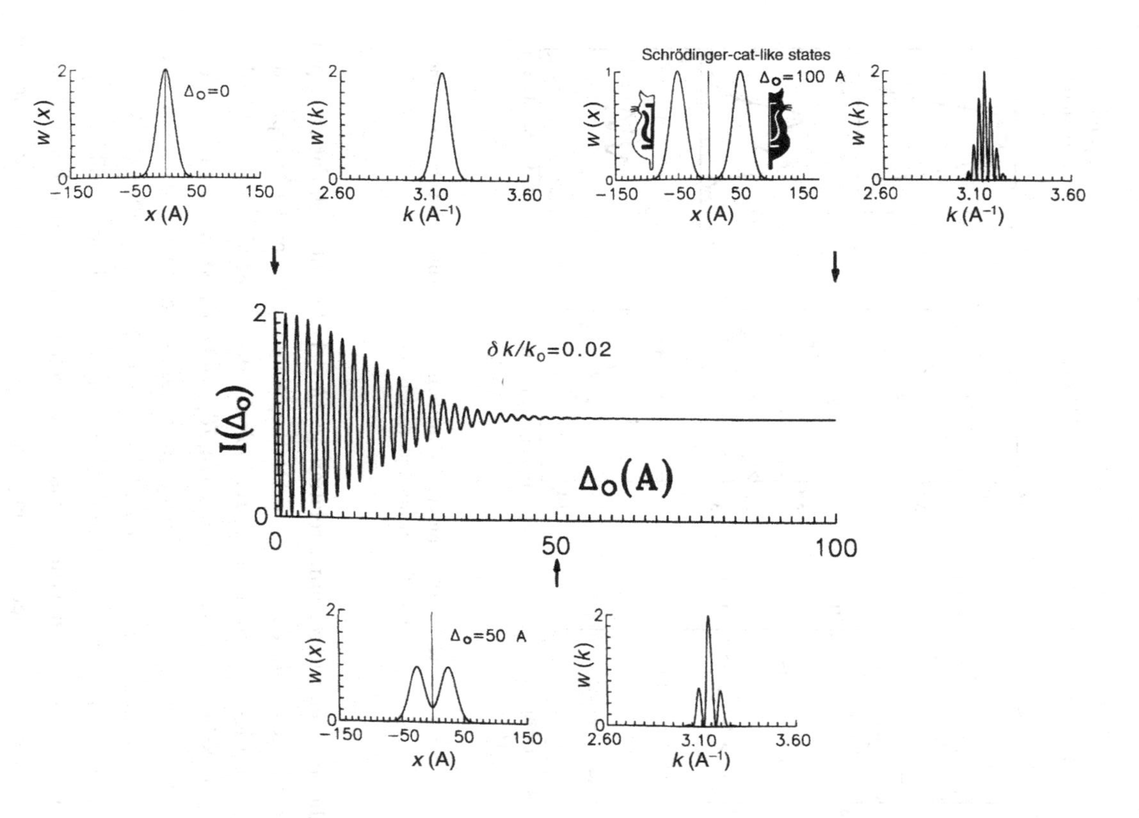

FIGURE 2.16 Wave packets and interference pattern in ordinary and momentum space [70].

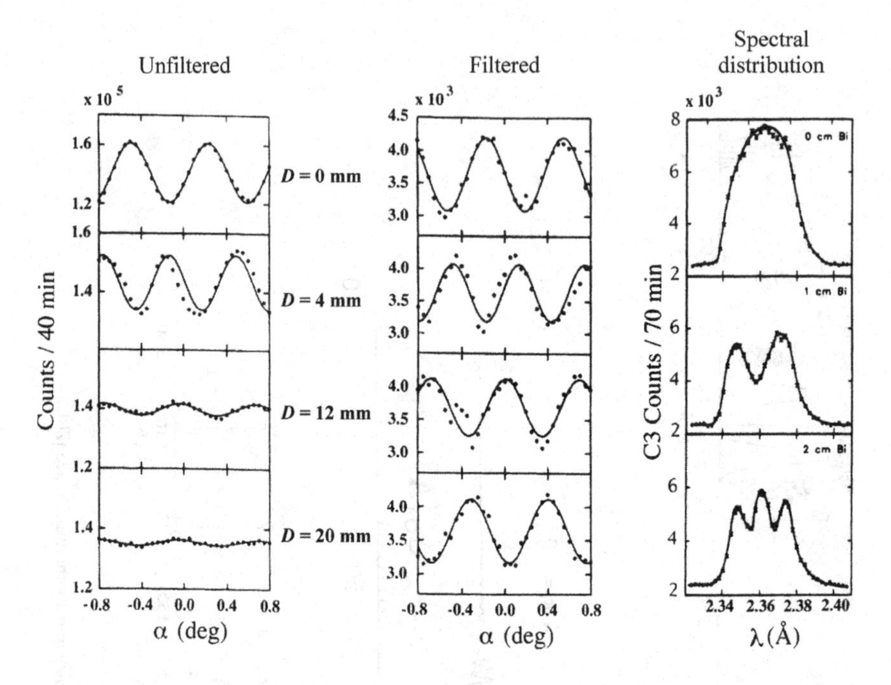

FIGURE 2.17 Measured interference pattern and momentum distributions in the case of momentum postselection [69].

Δ^c, the minimum uncertainty relation $\delta x/\delta k = 1/2$ is fulfilled. For an appropriately large displacement ($\Delta \gg \Delta^c$), the related state can be interpreted as a superposition state of two macroscopically distinguishable states, that is a stationary Schrödinger-cat like state [74, 76], but here first for massive particles. These states – separated in ordinary space and oscillating in momentum space – seem to be notoriously fragile and sensitive to dephasing effects [77–82].

Measurements of the wavelength spectrum were made with a silicon crystal with a rather narrow mosaic spread, which reflects in the parallel position a very narrow band of neutrons only ($\delta k'/k_0 \approx 0.0003$), causing an enhanced visibility at large phase shifts (Fig. 2.17). This feature shows that an interference pattern can be restored even behind the interferometer by means of a proper postselection procedure. In this case the overall beam does not show interference fringes any more and the wave packets originating from the two different beam paths do not overlap. The momentum distribution has been measured by scanning the analyser crystal through the Bragg-position. For different phase shifts the related results are shown in Fig. 2.17 as well [69]. These results clearly demonstrate that the predicted spectral modulation (Eq. (8.4)) appears when the interference fringes of the overall beam disappear. The modulation is somehow smeared out due to averaging processes across the beam which are due to various imperfections, unavoidably existing in any experimental arrangement. The contrast of the empty interferometer was 60%.

Each peak in the momentum distribution shown in Fig. 2.17 corresponds to a different number of phase shifts experienced by the neutrons of that wavelength band during its passage through the interferometer. In that sense, the minimum quantum entity of the incident wave packet becomes a new diverse entity representing different quantum states with distinguishable properties. This kind of labelling shows that constructive interference is restricted to a certain wavelength band only; a situation similar to that where new states have been created due to lattice diffraction inside the interferometer (Fig. 2.10) [21].

The new quantum states created behind the interferometer can be analysed with regard to their uncertainty properties. Analogies between a coherent state behaviour and a free but coherently coupled particle motion inside the interferometer have been addressed previously [16]. In such cases, the dynamical conjugate variables x and p minimize the uncertainty product with identical uncertainties $(\Delta x)^2 = (\Delta k)^2 = 1/2$ (in dimensionless units). Using $I_0(k)$ and $I_0(x)$ (Eqs. (8.5) and (8.6)) as distribution function we get a pronounced squeezing behaviour [69, 70, 74, 83–86]. One emphasizes that a single coherent state does not exhibit squeezing, but a state created by superposition of two coherent states can exhibit a considerable amount of squeezing. Thus highly nonclassical states are made by the power of the quantum mechanical superposition principle.

9. Request for postselection in EPR experiments

The previously discussed neutron experiments have shown us that phase space coupling persists even if the overlap in one parameter space does not exist any more. The stored information becomes exchanged between parameter spaces and can be measured by a proper experimental method. This has consequences for EPR experiments too. The entangled states (e.g. [87, 88]) of two photons

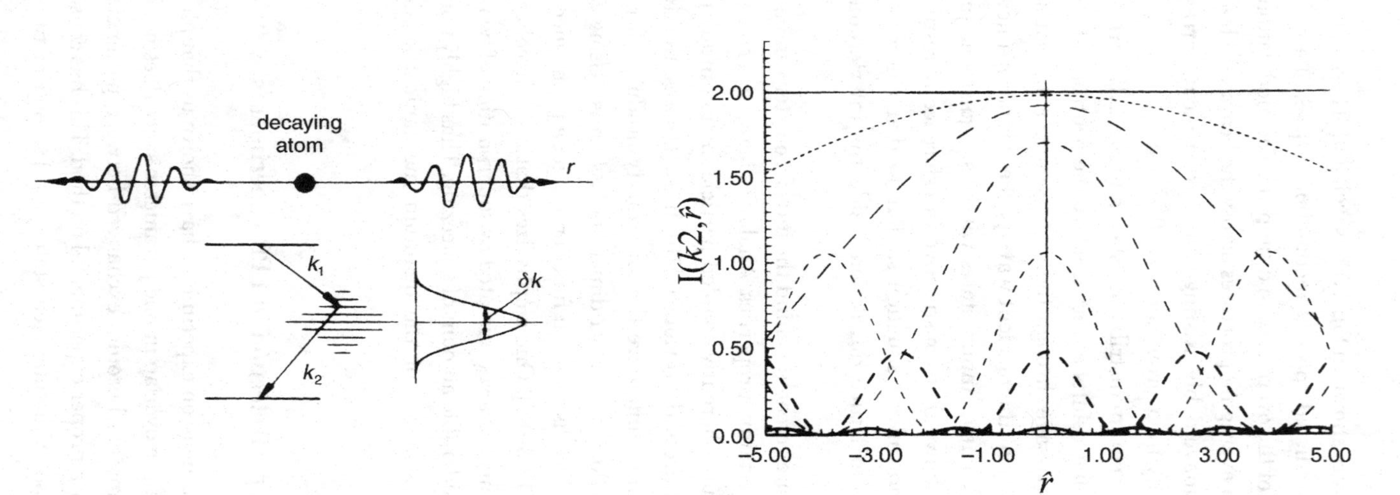

FIGURE 2.18 Scheme of an EPR-pair production by an atomic decay and the expected intensity modulation [70].

produced by an atomic decay cascade (Fig. 2.18),

$$\psi \propto |-k\rangle_1 |k\rangle_2 + |-k\rangle_2 |k\rangle_1 \tag{9.1}$$

are correlated due to the energy conservation of the transition,

$$k_1 + k_2 = k_{01} + k_{02} = \text{const.} \tag{9.2}$$

This produces a momentum and space dependent intensity distribution when the packet structure of the related wave functions is taken into account [70],

$$I(k_1, k_2, \vec{r}) = |\psi|^2 = 2|a(k_1)|^2 |a(k_2)|^2 \cdot (1 + \cos[2(k_2 - k_1)r]). \tag{9.3}$$

This shows a characteristic intensity modulation for each photon pair (Fig. 2.18) and indicates that individual $|k\rangle$ states remain interacting even at arbitrarily large spatial separation of the wave packets. For large distances ($r > (2\delta k)^{-1}$) the appearance of a momentum distribution modulation follows from Eq. (9.3) too [70]. If one of these photons is registered on one side its wave-function collapses, which instantaneously changes the wave function on the other side to $||k\rangle_2|^2$. This shows again that much more information can be gained than is usually extracted. Therefore, it is recommended that this experiment is repeated with a proper momentum resolution, which would show that the right and the left wave fields of the related momentum band (i.e. the partner photons) remain coupled even at arbitrarily large spatial separation of the overall wave-packets.

Related experiments will show that this coupling of the partner pairs of photons persists independently from its overall spatial separation. That indicates that locality should be treated in phase space rather than in ordinary space only. Here, too the required momentum resolution becomes more stringent when the packets become wider separated in ordinary space.

10. Discussion

All the results of the neutron interferometric experiments are well described by the formalism of quantum mechanics. According to the complementarity principle of the Copenhagen interpretation, the wave picture has to be used to describe the observed phenomena. The question how the well-defined particle properties of the neutron are transferred through the interferometer is not a meaningful one within this interpretation, but from the physical point of

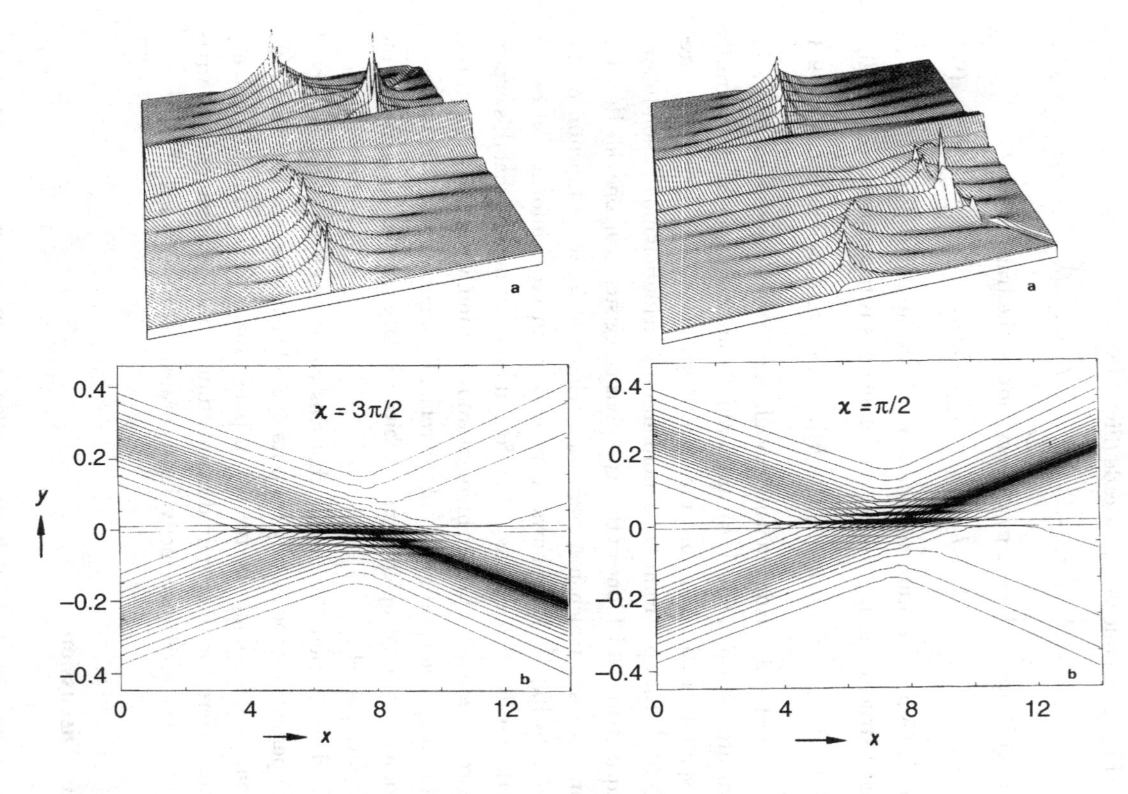

FIGURE 2.19 Quantum potential and beam trajectories at the place of beam superposition for a phase shift of $\chi = 3\pi/2$ (left) and $\chi = \pi/2$ (right) [67, 68].

view it should be an allowed one. Therefore, other interpretations should also be included in the discussion of such experiments. The particle picture can be preserved if pilot waves are postulated or if a quantum potential guides the particle to the predicted position. Related calculations have been performed for a simplified interferometer system [89, 90]. Unfortunately the results of these calculations are identical with the results of ordinary quantum mechanics and, therefore, to decide between both points of view remains an epistemological problem. The nonlocal quantum potential and the beam trajectories are shown in Fig. 2.19. The alternative view according to the wave picture is visualised in Fig. 2.20, where the position of the nodes of the superposed wave fields relative to the lattice points determine where the waves proceed behind the interferometer. Perhaps both pictures have to be combined and de Broglie's pilot waves have to be considered too; a view also supported by John Bell [91].

We have always tried to perform unbiased experiments and we do not wish to interfere with any epistemological interpretation of quantum mechanics. Perhaps in the future new proposals for experiments will be formulated, which will permit a unique decision between different interpretations. As an experimentalist, one appreciates the pioneering work of the founders of quantum mechanics, who created this basic theory with so little experimental evidence. Now we have much more direct evidence, even on a macroscopic scale, but, nevertheless, one notices that the interpretation of quantum mechanics goes beyond human intuition in certain cases. Only a few aspects of the experiments discussed before should be repeated. How can each neutron in the spin-superposition experiment be transferred from an initial pure state in the $|z\rangle$-direction into a pure state in the $|x\rangle$-direction behind the interferometer, if no spin turn occurs in one beam and a complete spin reversal occurs in the other beam path? How can every neutron have information about which beam to join behind the interferometer, when a slightly different energy exchange occurs in both beams insider the interferometer and the time constant of the beat effect is by many orders of magnitudes larger than the time of flight through the system? How can the interference pattern be influenced in a pulsed beam when the mean occupation number of a single pulse is of the order of 10^{-4} only? Various postselection experiments have shown that much more information is contained in the wave function than is usually measured. The experimental results also indicate that the question of locality and

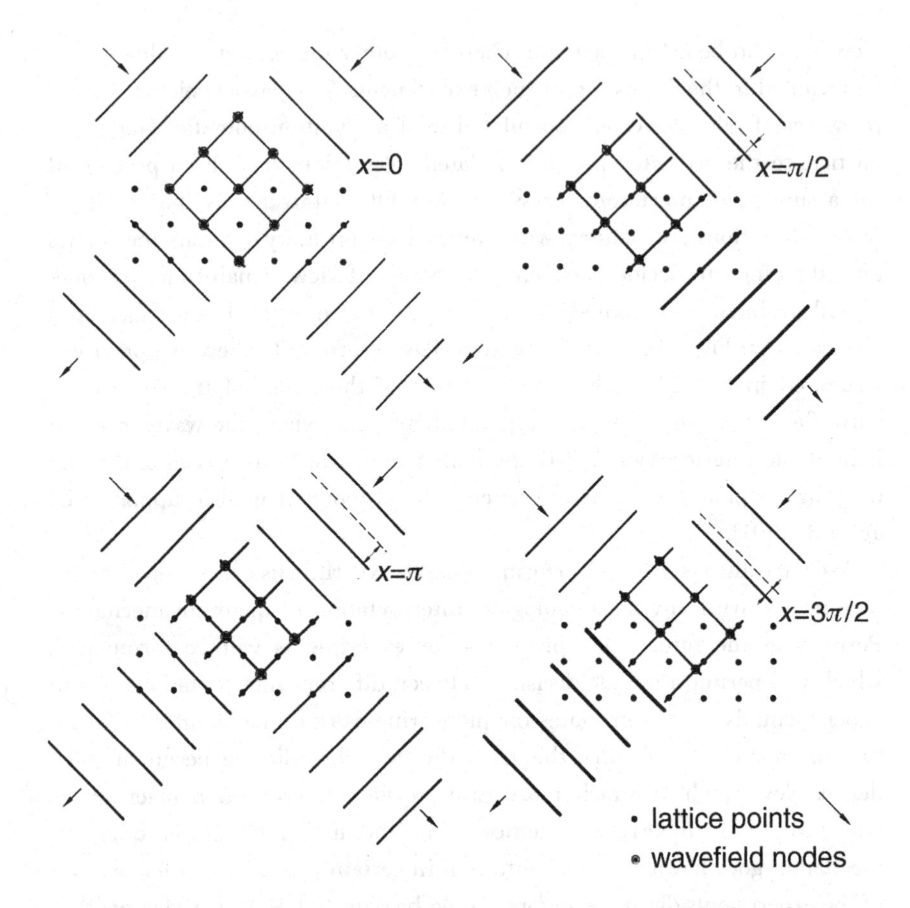

FIGURE 2.20 Nodes of the wave field and lattice points at the third interferometer plate. The relative position between the nodes of the wave field and the lattice points depends on the phase shift and determines the beams behind the interferometer.

nonlocality should be discussed in phase space rather than in ordinary space only. Coupling in phase space may exist even if the wave packets are well separated in ordinary space. This behaviour has been proven for neutron quantum states, but this phase space coupling may also help to understand EPR experiments in a new light. There are the speakable and unspeakable questions of quantum mechanics, often quoted by John Bell.

Experiments of our group and those which are related to fundamental physics problems have been discussed in this article. Several recent review articles can provide a broader view on the status of neutron interferometry [92–96].

ACKNOWLEDGMENTS

Most of the experimental results discussed in detail have been obtained by our Dortmund–Grenoble–Vienna interferometer group working at the high flux reactor in Grenoble, and some recent ones stem from our cooperation with the Columbia–Missouri group working at the MURR-reactor. The cooperation within these groups and especially the cooperation with colleagues from our Institute, which are cited in the references, is gratefully acknowledged.

REFERENCES

[1] H. Maier-Leibnitz and T. Springer, *Z. Physik* **167** (1962).
[2] R. Gaehler, J. Kalus and W. Mampe, *J. Phys.* **E13** (1980) 546.
[3] H. Rauch, W. Treimer and U. Bonse, *Phys. Lett.* **A47** (1974) 369.
[4] W. Bauspiess, U. Bonse, H. Rauch and W. Treimer, *Z. Physik* **271** (1974) 177.
[5] A.I. Ioffe, V.S. Zabiyankan and G.M. Drabkin, *Phys. Lett.* **111** (1985) 373.
[6] U. Bonse and M. Hart, *Appl. Phys. Lett.* **6** (1965) 155.
[7] H. Rauch and D. Petrascheck, *Neutron Diffraction*, H. Dachs, ed. (Springer Verlag, Berlin 1978) Chapter 9.
[8] V.F. Sears, *Can. J. Phys.* **56** (1978) 1261.
[9] W. Bauspiess, U. Bonse and W. Graeff, *J. Appl. Cryst.* **9** (1976) 68.
[10] D. Petrascheck, *Acta Phys. Austr.* **45** (1976) 217.
[11] M.L. Goldberger and F. Seitz, *Phys. Rev.* **71** (1947) 294.
[12] V.F. Sears, *Phys. Rep.* **82** (1982) 1.
[13] U. Bonse and H. Rauch (eds.), *Neutron Interferometry* (Clarendon Press, Oxford 1979).
[14] H. Rauch, E. Seidl, D. Tuppinger, D. Petrascheck and R. Scherm, *Z. Physik* **B69** (1987) 69.
[15] H. Rauch and J. Summhammer, *Phys. Lett.* **104A** (1984) 44.
[16] J. Summhammer, H. Rauch and D. Tuppinger, *Phys. Rev.* **A36** (1987) 4447.
[17] H. Rauch, J. Summhammer, M. Zawisky and E. Jericha, *Phys. Rev.* **A42** (1990) 3726.
[18] M. Namiki and S. Pascazio, *Phys. Lett.* **147** (1990) 430.
[19] J. Bell, *Physics* **1** (1965) 195.
[20] D. Home and F. Selleri, *Revista del Nuovo Cim.* **14** (1991) 1.
[21] H. Rauch and J. Summhammer, *Phys. Rev.* **A46** (1992) 7284.

[22] H. Rauch, *Proc. 3rd Int. Symp. Found. Quantum Mechanics*, Shun-ichi Kobayashi *et al.*, eds. (Phys. Soc. Japan, 1989), p. 3.

[23] H. Rauch, J. Summhammer, M. Zawisky and E. Jericha, *Phys. Rev.* **A42** (1990) 3726.

[24] P. Mittelstaed, *Proc. 3rd Int. Symp. Found. Quantum Physics*, Tokyo, 1989, p. 153.

[25] B.-G. Englert, *Phys. Rev. Lett.* **77** (1996) 2154.

[26] Y. Aharonov and L. Susskind, *Phys. Rev.* **158** (1967) 1237.

[27] H.J. Bernstein, *Phys. Rev. Lett.* **18** (1967) 1102.

[28] G. Eder and A. Zeilinger, *Nuovo Cim.* **34B** (1976) 76.

[29] H. Rauch, A. Zeilinger, G. Badurek, A. Wilfing, W. Bauspiess and U. Bonse, *Phys. Lett.* **A54** (1975) 425.

[30] S.A. Werner, R. Colella, A.W. Overhauser and C.F. Eagen, *Phys. Rev. Lett.* **35** (1975) 1053.

[31] A. Zeilinger, *Nature* **294** (1981) 544.

[32] H.J. Bernstein, *Nature* **315** (1985) 42.

[33] H. Rauch, A. Wilfing, W. Bauspiess and U. Bonse, *Z. Physik* **B29** (1978) 281.

[34] M.V. Berry, *Proc. R. Soc. London* **A392** (1984) 45.

[35] Y. Aharonov and J. Anandan, *Phys. Rev. Lett.* **58** (1987) 1593.

[36] A.G. Wagh, V.C. Rakhecha, J. Summhammer, G. Badurek, H. Weinfurter, B.M. Allman, H. Kaiser, K. Hamacher, D.L. Jacobson and S.A. Werner, *Phys. Rev. Lett.* **78** (1997) 755.

[37] Y. Hasegawa, M. Zawisky, H. Rauch and A.I. Ioffe, *Phys. Rev.* **A53** (1996) 3486.

[38] J. Summhammer, G. Badurek, H. Rauch, U. Kischko and A. Zeilinger, *Phys. Rev.* **A27** (1983) 2523.

[39] G. Badurek, H. Rauch and J. Summhammer, *Phys. Rev. Lett.* **51** (1983) 1015.

[40] A. Zeilinger, in [13], p. 241.

[41] E.P. Wigner, *Am. J. Phys.* **31** (1963) 6.

[42] B. Alefeld, G. Badurek and H. Rauch, *Z. Physik* **B41** (1981) 231.

[43] G.M. Drabkin and R.A. Zhitnikov, *Sov. Phys. JETP* **11** (1960) 729.

[44] F. Bloch and A. Siegert, *Phys. Rev.* **57** (1940) 522.

[45] H. Kendrick, J.S. King, S.A. Werner and A. Arott, *Nucl. Instr. Meth.* **79** (1970) 82.

[46] P. Carruthers and M.M. Nieto, *Rev. Mod. Phys.* **40** (1968) 411.

[47] R. Jackiw, *J. Math. Phys.* **9** (1968) 339.

[48] R.J. Glauber, *Phys. Rev.* **131** (1963) 2766.

[49] M.O. Scully and H. Walther, *Phys. Rev.* **A39** (1989) 5229.

[50] C. Dewdney, P. Gueret, A. Kyprianidis and J.P. Vigier, *Phys. Lett.* **102A** (1984) 291.

[51] J.P. Vigier, *Pramana* **25** (1985) 397.
[52] G. Badurek, H. Rauch and D. Tuppinger, *Proc. Int. Conf. New Techniques and Ideas in Quantum Measurement Theory* (New York, Jan. 1986) (New York Academy of Science, 1986) 133.
[53] G. Badurek, H. Rauch and D. Tuppinger, *Phys. Rev.* **A34** (1986) 2600.
[54] H. Rauch and J.P. Vigier, *Phys. Lett.* **151** (1990) 269.
[55] H. Rauch, *Symp. Found. Modern Physics*, P. Lahti, P. Mittelstaedt, eds., Joensuu, 1990 (World Sci. Publ.) p. 347.
[56] B.D. Hosephson, *Rev. Mod. Phys.* **46** (1974) 251.
[57] J. Summhammer, K.A. Hamacher, H. Kaiser, H. Weinfurter, D.L. Jacobson and S.A. Werner, *Phys. Rev. Lett.* **75** (1995) 3206.
[58] F. Mezei (ed.), *Neutron Spin Echo*, Lect. Notes in Physics **128** (Springer, 1980).
[59] G. Badurek, H. Rauch and A. Zeilinger, in [58], p. 136.
[60] H. Rauch, in [13], p. 161.
[61] H. Kaiser, S.A. Werner and E.A. George, *Phys. Rev. Lett.* **50** (1983) 560.
[62] R. Clothier, H. Kaiser, S.A. Werner, H. Rauch and H. Wölwitsch, *Phys. Rev.* **A44** (1990) 5357.
[63] A. Messiah, *Quantum Mechanics* (North-Holland, Amsterdam, 1965).
[64] M. Moshinsky, *Phys. Rev.* **88** (1952) 625.
[65] R. Gähler and R. Golub, *Z. Physik* **B56** (1984) 5.
[66] H. Rauch, H. Wölwitsch, R. Clothier, H. Kaiser and S.A. Werner, *Phys. Rev.* **A46** (1992) 49.
[67] S.A. Werner, R. Clothier, H. Kaiser, H. Rauch and H. Wölwitsch, *Phys. Rev. Lett.* **67** (1991) 683.
[68] H. Kaiser, R. Clothier, S.A. Werner, H. Rauch and H. Wölwitsch, *Phys. Rev.* **A45** (1992) 31.
[69] D.L. Jacobson, S.A. Werner and H. Rauch, *Phys. Rev.* **A49** (1994) 3196.
[70] H. Rauch, *Phys. Lett.* **A173** (1993) 240.
[71] W. Schleich and J.A. Wheeler, *Nature* **326** (1987) 574.
[72] W. Schleich, D.F. Walls and J.A. Wheeler, *Phys. Rev.* **A38** (1988) 1177.
[73] J.-M. Levy-Leblond and F. Balibar, *Quantics* (North-Holland, 1990).
[74] W. Schleich, M. Pernigo and Fam Le Kien, *Phys. Rev.* **A44** (1991) 2172.
[75] A. Legett, *Proc. Found. Quantum Mechanics*, S. Kamefuchi, ed. (Phys. Soc. Japan, 1984), p. 74.
[76] B. Yurke, W. Schleich and D.F. Walls, *Phys. Rev.* **A42** (1990) 1703.
[77] D.F. Walls and G.J. Milburn, *Phys. Rev.* **A31** (1985) 2403.
[78] R.J. Glauber, New techniques and ideas, in: *Quantum Measurement Theory*, D.M. Greenberger, ed. (N. Y. Acad. Sci., 1986) p. 336.
[79] M. Namiki and S. Pascazio, *Phys. Rev.* **A44** (1991) 39.
[80] H. Zurek, *Physics Today* (Oct. 1991) p. 36.
[81] P. Blanchard and A. Jadczyk, *Phys. Lett.* **A175** (1993) 157.

[82] V. Buzek, C.H. Keitel and P.L. Knight, *Phys. Rev.* **A51** (1995) 2594.
[83] D.F. Walls, *Nature* **306** (1983) 141.
[84] S.L. Braunstein and R.I. McLachlan, *Phys. Rev.* **A35** (1987) 1659.
[85] R. Loudon and P. L. Knight, *J. Mod. Opt.* **34** (1987) 709.
[86] J. Janski and A.V. Vinogradov, *Phys. Rev. Lett.* **64** (1990) 2771.
[87] A. Einstein, B. Podolsky and N. Rosen, *Phys. Rev.* **47** (1935) 777.
[88] D.M. Greenberger, M.A. Horne and A. Zeilinger, in: *Bell's Theorem, Quantum Theory and Conceptions of the Universe*, M. Kafatos, ed. (Kluwer Publ., 1989) p. 69.
[89] C. Dewdney, *Phys. Lett.* **109A** (1985) 377.
[90] C. Dewdney, P.R. Holland and A. Kyprianidis, *Phys. Lett.* **A119** (1986) 259.
[91] J. Bell, *Plenary Lecture, 100th Birthday Schrödinger Symposium*, Vienna 1988.
[92] A.G. Klein and S.A. Werner, *Rep. Progr. Phys.* **46** (1983) 259.
[93] D. Greenberger, *Rev. Mod. Phys.* **55** (1983) 875.
[94] H. Rauch, *Contemp. Phys.* **27** (1986) 345.
[95] S.A. Werner and A.G. Klein, in *Methods of Experimental Physics 23*, Part A, 259 (Academic Press, 1986).
[96] V.F. Sears, *Neutron Optics* (Oxford University Press, 1989).

3 Testing Bell's inequalities

ALAIN ASPECT

This chapter, written in 1991, shows that the results of precise tests to explore the validity of quantum mechanics by applying Bell's theorem (that some quantum mechanical predictions cannot be mimicked by any local realistic model in the spirit of Einstein's ideas) agree with the quantum mechanical predictions.

It is a great privilege to have the opportunity to recount the great influence John Bell had on my life as a physicist. Testing Bell's inequalities was more than a run-of-the-mill experiment. Indeed, when I read the paper 'On the Einstein–Podolsky–Rosen (EPR) paradox [1], I found it extremely clear and completely convincing, but there was something special about this paper: it led to two contradictory conclusions. The first part showed that EPR correlations predicted by quantum mechanics are so strong that one can hardly avoid the conclusion that quantum mechanics should be completed by some supplementary parameters (the so-called 'hidden variables'). But the second part, elaborating on this result, demonstrated that the hidden-variables description in fact contradicts some predictions of quantum mechanics, which is to say the two theories predict different results. In the face of these two perfectly convincing and contradictory results, there is only one way out: ask Nature how it works.

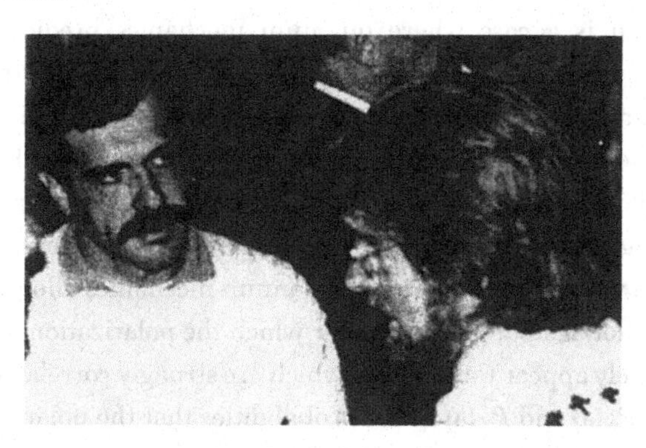

FIGURE 3.1 A. Aspect (on the left) with John Bell in about 1985 in Paris.

The big surprise was the realization that, at the end of the sixties, there was no experimental result to answer the question. The contradiction discovered by John Bell is so subtle that it appears only in very peculiar situations that had not been investigated: it was therefore necessary to design and build specific experiments.

Bell's theorem

1. Hidden variables

The reasoning behind Bell's theorem deals with correlations between events, each of which appears to be random. Such correlations may arise outside physics. Take, for instance, the occurrence of some well-defined disease and let us assume that biologists have observed its development in 50% of the population aged 20, and its absence in the remaining half. Now, on investigating specific pairs of (true) twin brothers, they find a perfect correlation between the outcomes: if one brother (or sister) is affected, the other is also found to be inflicted with the disease; but if one member of the pair has not developed the disease, then the other is also unaffected. In face of such a perfect correlation for twin brothers, the biologists will certainly conclude that the disease has a genetic origin. They may invoke a simple scenario: at the first step of conception of the embryo, a (random) genetic process produced a chromosome sequence – one which is responsible for the occurrence, or absence, of the disease – that has been duplicated and given to both brothers.

An EPR situation is a case where quantum mechanics predicts strong correlations of this type. Consider, for instance, the situation illustrated in Fig. 3.2 where a source emits a pair of photons ν_1 and ν_2 travelling in opposite directions. Each photon impinges onto polarizers which measure the linear polarization along both of two directions (**a** or **b**) determined by the orientation of the corresponding polarizer. There are two possible outcomes for each measurement and these we can label $+$ and $-$. Quantum mechanics allows for the existence of a two-photon state (EPR state) for which the polarization measurements taken separately appear random but which are strongly correlated. More precisely, denoting $P_+(\mathbf{a})$ and $P_-(\mathbf{a})$ as the probabilities that the polarization of ν_1 along **a** is found equal to $+$ or $-$, these probabilities are predicted to be equal

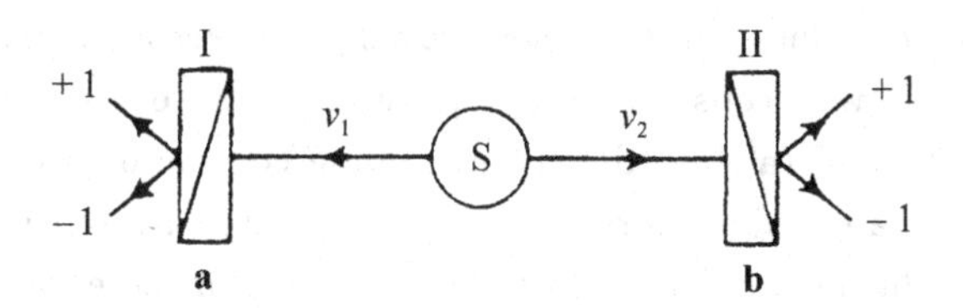

FIGURE 3.2 Einstein–Podolsky–Rosen Gedankenexperiment with photons. The source S emits pairs of photons ν which are analysed in polarization in two directions (**a** and **b**). In an EPR situation, the results of the measurement of polarizations are found to be strongly correlated.

to 0.5; similarly the probabilities $P_+(\mathbf{b})$ and $P_-(\mathbf{b})$ for photon ν_2 are equal to 0.5 and independent of the orientation **b**.

On the other hand, the joint probability $P_{++}(\mathbf{a}, \mathbf{b})$ for observing + for both photons is equal to $0.5 \cos^2(\mathbf{a}, \mathbf{b})$. In the case of parallel polarizers $[(\mathbf{a}, \mathbf{b}) = 0]$, this joint probability is $P_{++}(0) = 0.5$; similarly, $P_{--}(0) = 0.5$, while $P_{-+}(0)$ and $P_{+-}(0)$ are zero. The results for the two photons of the same pair are thus always identical, both + or both −, i.e., they are completely correlated. The situation is thus exactly analogous to the case for the twin brothers, and it seems natural to link this correlation to some common property of the two photons of a pair, analogous to the common genome of the two twin brothers. This common property changes from pair to pair, which accounts for the random character of the single events.

The above reasoning constitutes the first part of John Bell's paper. A natural generalization of the EPR reasoning, it leads to the conclusion that quantum mechanics is not a complete description of physical reality. Indeed, invoking some common property which changes from pair to pair, we claim that the complete description of a pair must include something in addition to the state vector which is the same for all pairs. This something can be called *supplementary parameters*, or *hidden variables*. At this stage, these hidden variables are supposed to be able to render an account of the correlations between both measurements, for any set $(\mathbf{a}, \mathbf{b})$ of orientations.

2. Inequalities

The second part of Bell's reasoning starts from this requirement for hidden variables. Assuming their existence and some very natural properties, one can show that the expected correlations, for the joint measurements above, cannot take any set of values, but that they are subject to certain constraints. More precisely, if we consider four possible sets of orientations $[(\mathbf{a}, \mathbf{b}), (\mathbf{a}, \mathbf{b}'), (\mathbf{a}', \mathbf{b})$ and $(\mathbf{a}', \mathbf{b}')]$, the corresponding correlation coefficients (which measure the amount of correlation) are restricted by the so-called Bell's inequalities, which state that a given combination S of these four coefficients is between -2 and $+2$ for any 'reasonable' hidden-variable theory [2].

Now comes the crucial point: there exists a set of orientations for which the quantity S predicted by quantum mechanics, in the EPR situation presented above, is equal to 2.8, i.e., it violates Bell's inequalities. The hidden-variables theories envisaged above are then unable to render an account of the EPR correlations predicted by quantum mechanics (these quantum mechanical correlations are not as easy to understand as the common medical fate of twin brothers).

In the face of this contradiction, John Bell made clear the reasonable properties that he had assumed for the hidden-variables models. The essential assumption, absolutely necessary to resolve the conflict, is locality: this assumption states that the result of a measurement by a polarizer cannot be directly influenced by the choice of the orientation of the other, remotely located, polarizer. This assumption indeed sounds very reasonable. Moreover, it can be considered to be a consequence of Einstein's causality, by considering an experiment in which the settings of the polarizers can be changed at random in a time which is short compared to the time light takes to propagate between the two polarizers.

From the many papers that followed Bell's paper, we will extract the conclusion that Bell's inequalities apply to a wider class of theories than local hidden-variable theories. Any theory in which each photon has a 'physical reality' localized in space-time, determining the outcome of the corresponding measurement, will lead to inequalities that sometimes conflict with quantum mechanics. Bell's theorem can thus be phrased in the following way: *some quantum mechanical predictions (EPR correlations) cannot be mimicked by any local realistic model in the spirit of Einstein's ideas.*

First experiments [3]

When physicists realized the wide generality of Bell's theorem, they met with a great surprise: at the end of the 1960s, there was no experimental result available for testing Bell's inequalities *versus* quantum mechanics. Moreover, in the case where a conflict is predicted (as above) one finds that taking into account the inefficiencies of a real experiment usually reduces the degree of correlation predicted by quantum mechanics so that there is no longer any conflict. The possibility then arose that the conflict with Bell's inequalities may indicate a place to look for a limit to the validity of quantum mechanics.

It was therefore tempting to perform a sensitive experiment for a situation where quantum mechanics predicts a conflict with Bell's inequalities. In order to have such a situation, several conditions must be fulfilled:

- the creation of a pair of systems in a non-factorizable quantum state of the EPR type;
- the ability to perform two-valued measurements on each system;
- the disposal of an adjustable parameter for these measurements so that different values of this parameter correspond to non-commuting observables.

The first experimental test was based on pairs of γ photons produced in the annihilation of positronium. This would be an ideal system, except for the fact that no polarizer exists capable of making a two-valued measurement: polarization must be inferred from a Compton scattering using calculations relying on quantum theory. The test is thus indirect and somewhat circular. The first measurements gave contradictory results, but by the mid-1970s clear agreement with the quantum mechanical predictions was established.

An experiment based on pairs of protons obtained by scattering had the same problem (no polarizers). It also gave a result in agreement with quantum mechanics.

The system best able to fulfil the above conditions comprises pairs of visible photons produced in well-chosen atomic radiative cascades. As a matter of fact, for visible light there exist polarizers, e.g. based on birefringent crystals, with two output channels and an adjustable orientation. The first three experiments, carried out in the early 1970s, gave a relatively small signal and some results were contradictory. By introducing a laser to excite the cascade, the fourth experiment gave a convincing result in agreement with quantum mechanics. For practical reasons, all these early experiments used only one-channel polarizers, so once again the comparison of the experimental results with Bell's inequalities was indirect and relied on supplementary assumptions. However, they had given convincing indications in favour of quantum mechanics and they opened the way to second-generation experiments.

Closer to the Gedankenexperiment

Thanks to the progress in lasers, we could design and build in the late 1970s [4] a much more efficient source of pairs of EPR photons correlated in polarization. We used the same radiative cascade in calcium-40 as employed in the first experiment by Clauser and Freedman [2], but now we could selectively excite the upper level of the cascade with two-photon absorption. As a consequence, the light emitted by our source was very pure, encompassing only photons of the desired pairs. Very important also was the very high emission rate which allowed us to achieve a 1% statistical accuracy for joint detection within only one hundred seconds (a similar level of accuracy required hours in the previous experiments).

A first experiment based on the same scheme as the previous ones (with one-channel polarizers) gave a clear-cut result in agreement with quantum mechanics. Meanwhile, we had obtained (from the Philips Research Laboratory) two-channel polarizers based on multi-dielectric coatings. Figure 3.3

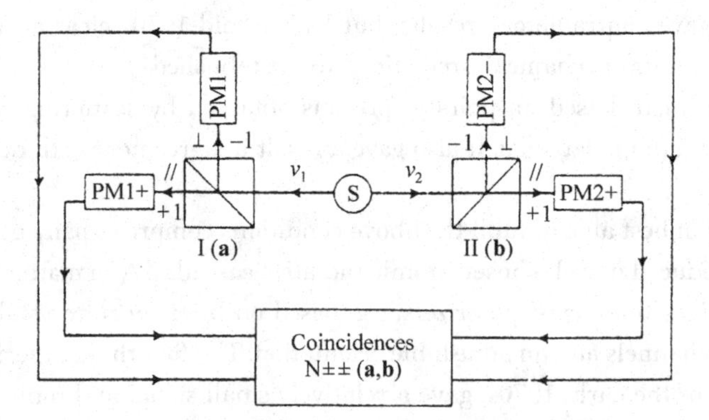

FIGURE 3.3 The experiment with two-channel polarizers and photomultiplier (PM) counters. Using four-fold coincidence systems we obtain directly the polarization correlation coefficient of photons ν emitted by the source S in the set of orientations (**a**, **b**). Note the close similarity with Fig. 3.2.

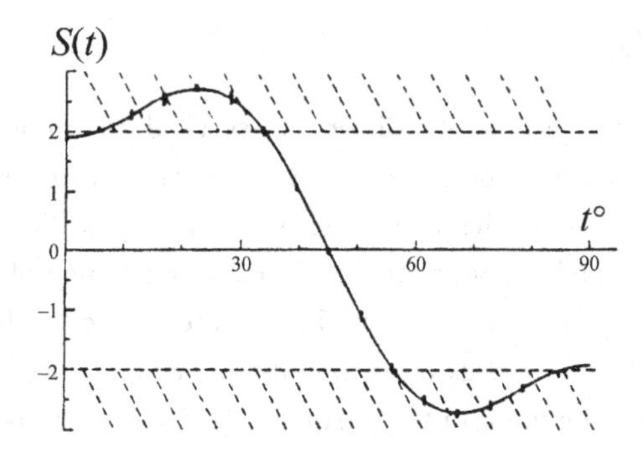

FIGURE 3.4 The results of the experiment of Fig. 3.3. The quantity S is a function of the correlation coefficients and should lie between -2 and 2 according to Bell's inequalities. The solid curve is the prediction of quantum mechanics taking into account inefficiencies of the apparatus.

shows the experimental set-up, which closely resembled the ideal one of Fig. 3.2. Using the four-fold coincidence system it was possible to monitor simultaneously the four coincidence rates corresponding to the $+$ and $-$ results. This yields directly, without any auxiliary calibration, the joint detection probabilities in a given set of orientations $(\mathbf{a}, \mathbf{b})$ from which we derive the correlation coefficient. By repeating the measurement in different orientations we can test directly Bell's inequalities.

The results are shown in Fig. 3.4. We have plotted, as a function of the angle t, the quantity S which is subject to the Bell's inequalities:

$$-2 \leq S \leq 2.$$

There are obviously angles for which one of the Bell's inequalities is violated. The maximum violation corresponds to a value

$$S = 2.70 \pm 0.015,$$

that is to say a violation by more than 40 standard deviations. In spite of its close resemblance to the ideal experiment, the actual experiment suffers from one remaining problem: owing to the limited efficiency of the photon detectors, a comparison with Bell's inequalities requires the assumption that the detected photons constitute a faithful sample. Nevertheless, the result in favour of quantum mechanics and against local hidden-variables theories is very convincing.

Testing locality

As already emphasized, the locality condition is essential to obtain Bell's inequalities. But, as stressed by John Bell, in an experiment of the type described above, 'the settings of the instruments are made sufficiently in advance to allow them to reach some mutual rapport by exchange of signals with velocity less than or equal to that of light', in which case the locality condition does not apply [1]. We have thus tried to realize a scheme 'in which the settings are changed during the flight of the particles', so that locality be a consequence of Einstein's causality (no interaction can propagate faster than light).

We have in fact only partially realized this programme. First, we do not in practice change the setting of a polarizer but we instead replace each polarizer with a system involving a switch that is able to redirect the light towards one of two polarizers in two different orientations, as shown in Fig. 3.5. The time between two changes is 10 nanoseconds, shorter than the time of flight of the photons (20 nanoseconds, corresponding to 6 metres). Unfortunately, the switches (based on the interaction with an acoustic standing wave) did not work at random but periodically. This is far from ideal, even if the two switches are driven by independent generators.

Owing to the complication of the systems, the signal was smaller than in the static experiment of Fig. 3.2, and the results were not as precise. We

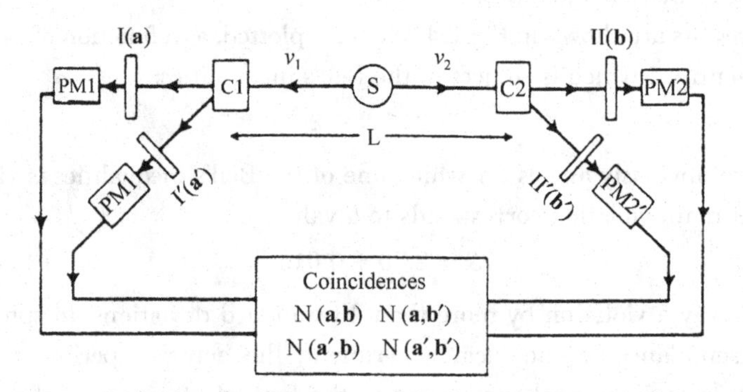

FIGURE 3.5 The experiment with optical switches. The switch C1, with the two polarizers I and I′ is equivalent to a single polarizer changed from orientation a to orientation a′. The time between changes is shorter than the time of flight of the light. As before, the source S emits pairs of photons ν and photomultipliers (PM) detect the photons.

nevertheless obtained a significant violation of Bell's inequalities by five standard deviations, and a good agreement with quantum mechanics. The level of confidence of this result is not as high as in the earlier experiments, so it would be very interesting to perform another experiment of the same type. This might be done with one of the new sources of pairs of photons produced in the parametric down-conversion of photons [5] which should eventually give better results than our source.

The non-locality heritage

Let us assume that quantum mechanics will also work in ideal experiments with no inefficiencies present. In the words of John Bell [6]: 'It is difficult for me to believe that quantum mechanics, working very well for currently practical set-ups, will nevertheless fail badly with improvements in counter efficiency and other factors . . .'. What can we conclude? We cannot do better than let John Bell explain possible attitudes:

> There are influences going faster than light, even if we cannot control them for practical telegraphy. Einstein local causality fails, and we must live with this.
>
> The orientations **a** and **b** are not independently variable as we supposed. Whether apparently chosen by apparently independent radioactive devices, or by apparently separate Swiss National Lottery machines, or even by different apparently free-willed experimental physicists, they are in fact correlated with the same causal factors as the A and B (the outcomes of the measurements). Then Einstein causality can survive. But apparently separate parts of the world become deeply entangled, and our apparent free will is entangled with them.
>
> The whole analysis can be ignored. The lesson of quantum mechanics is not to look behind the predictions of the formalism. As for the correlations, well, that's quantum mechanics.

John Bell repeatedly made it clear that the last attitude was not his. To renounce raising difficult questions would not have been acceptable to him. The first was apparently his favourite; like the second, it leaves us with a world, the various parts of which may be deeply entangled. After John Bell, we can no longer ignore that the quantum physical reality is somewhat non-local.

REFERENCES

[1] J.S. Bell, *Physics* **1** (1964) 195. Like the other papers by J.S. Bell on the foundations of quantum mechanics, this paper may be found more easily in J.S. Bell, *Speakable and Unspeakable in Quantum Mechanics* (Cambridge University Press, Cambridge, U.K., 1987).

[2] Instead of the original form of the Bell's inequalities, we use this form, more convenient for experimental tests, first derived by J.F. Clauser, M.A. Horne, A. Shimony and R.A. Holt, *Phys. Rev. Lett.* **37** (1965), 465.

[3] Excellent reviews of the situation in 1978 are given in: J.F. Clauser and A. Shimony, *Rep. Progr. Phys.* **41** (1978) 1881; F.M. Pipkin, *Adv. in Atomic & Molecular Physics* **14** (1978) 281. See also an interesting more recent experiment: W. Perrie *et al.*, *Phys. Rev. Lett.* **54** (1985) 1790.

[4] These experiments were carried out in 1978–82 at the Institut d'Optique d'Orsay, Université de Paris Sud, with P. Grangier, G. Roger and J. Dalibard. See e.g.: *Atomic Physics* **8** (1983) 103.

[5] Y.H. Shih and C.O. Alley, *Phys. Rev. Lett.* **61** (1988) 2921; Z.Y. Ou and L. Mandel, *Phys. Rev. Lett.* **61** (1988) 50; J.G. Rarity and P.R. Tapster, *Phys. Rev. Lett.* **64** (1990) 2995.

[6] J.S. Bell, *Comm. on Molecular Physics* **9** (1980) 121.

4 Beyond conventional quantum mechanics

GIANCARLO GHIRARDI

1. Introduction

We review some recent attempts to overcome the conceptual difficulties which one encounters when trying to interpret quantum mechanics as giving a complete, objective and unified description of natural phenomena. Concerning the above specifications, we recall that the completeness requirement amounts to pretending that the state vector represents the most accurate possible characterization of the state of a physical system and demands disregarding the eventuality that extra (possibly hidden) parameters are necessary. The term objective refers to the refusal to attribute a peculiar role to the observer and to the act of observation, while the term unified refers to the requirement that all physical processes, the microscopic and the macroscopic ones, as well as the micro–macro interactions which characterize the measurement processes, are governed by the same basic dynamics holding for the 'elementary constituents' of physical systems. The specification 'beyond conventional' appearing in the title refers to the fact that the attempts considered, while requiring, as already stated, that the wave function gives a complete description of physical systems, plainly accept that [1] '*Schrödinger's equation is not always right*'* and correspondingly are based on a modification of the linear and deterministic evolution law which is characteristic of the standard formulation of quantum mechanics.

John Bell took these proposals very seriously. His having been [2] *intrigued* by them has meant, for all people involved in this program, much more than words can say. His continuous interest, his stimulating remarks and his passionate defence played, as we will show below, an important role in the development of this line of research.

* Throughout this chapter sentences in italics and quotes are taken from writings of J.S. Bell, unless stated otherwise.

We will not try to analyze the pros and cons of these 'beyond conventional quantum mechanics' attempts with respect to other proposed solutions to the difficulties of the theory. Our attitude is simply that, since they represent a line which nobody would have considered viable up to few years ago and which has been proved to be perfectly workable and to lead to the possibility of taking a macro-objective position about natural phenomena, they deserve some attention and have to be taken into account in discussing the conceptually fundamental issues raised by quantum mechanics.

We will, instead, try to stress to which extent the elaboration of the models we will discuss below has allowed us to meet the requirements that J. Bell always put forward for 'a serious theory'. The deep motivations for advancing such requirements and for his lucid struggle against what he considered the complacency of [3] '*the good unproblematic formulations*' of the theory are well known: the sake of physical precision, the refusal of pragmatic '*FAPP (for all practical purposes)*' attitudes, an honest and rigourous position about the aims of science which cannot be confined to speak '*about our piddling laboratory operations*', thus '*betraying the great enterprise*'. Nobody took more seriously than him the necessity of this pursuit of reality. We want to stress that it was just this desire to understand reality and in no way some metaphysical prejudice* that led him to repeatedly and passionately assert that professional scientists should be able to do better and to achieve [3] an# '*exact formulation of some serious part of quantum mechanics*'.

2. The program

In this section we will review the investigations along the 'beyond conventional quantum mechanics' line of thought which have led to the elaboration of

* Actually J. Bell always had an absolutely open minded attitude about science and was ready to grasp all lessons coming from our investigations about nature, provided he could see that they might fit into a coherent worldview. This emerges clearly from the consideration of the revolutionary conceptual change of attitude which was necessary to derive his famous inequality and from almost all his writings, a significant example being given by the sentence appearing in one of his last papers [3]: '*suppose that ... we find an unmovable finger obstinately pointing outside the subject, to the mind of the observer, to the Hindu scriptures, to God, or even only Gravitation? Would not that be very very interesting?*'.

To properly appreciate some of the quotations in this chapter it is important to recall that J. Bell has always stressed that by 'exact' he does not mean 'exactly true'. To pretend that a theory be true would seem to him to lack in humility. Analogously it has to be kept in mind that by 'serious' he means covering [3] '*some substantial fragment of physics*'.

precise non-relativistic models of dynamical state vector reduction. The relativistic case will be considered in Section 3.

2.1. The motivations

We summarize the situation one has to face when trying to build up a coherent worldview based on the quantum picture of natural phenomena. The critical investigations which led to the elaboration of the quantum scheme implied the recognition that the states of microscopic physical systems obey the superposition principle. The linear nature of the theory entails the conceptual impossibility of attributing to physical systems too many 'possessed properties' and, correspondingly, the [1] '*indefiniteness, the waviness*' of microphenomena. As a simple example we recall that if consideration is given to the following state of a single microscopic system:

$$|\Psi\rangle = N[|h\rangle + |t\rangle], \tag{2.1}$$

(N is a normalization factor), where the two states $|h\rangle$ and $|t\rangle$ have the coordinate representation

$$\langle \mathbf{q}|h\rangle = [\pi\delta^2]^{-3/4} \exp\left[\frac{(\mathbf{q}-\mathbf{h})^2}{2\delta^2}\right], \quad \langle \mathbf{q}|t\rangle = [\pi\delta^2]^{-3/4} \exp\left[\frac{(\mathbf{q}-\mathbf{t})^2}{2\delta^2}\right], \tag{2.2}$$

and $\mathbf{h}$ (here) and $\mathbf{t}$ (there) denote two far away positions in space such that $\delta \ll |\mathbf{h}-\mathbf{t}|$, then, as is well known, one cannot legitimately state that the system possesses the property of being EITHER here OR there, while an (almost) precise location can be attributed to the system when it is in one of the two states appearing in the superposition (2.1).

In going from elementary to composite systems a new feature, 'entanglement', emerges. It implies the loss of individuality of the constituents and the impossibility of attributing to them (in the most general case) any property at all; nevertheless it allows the existence of strict correlations for the outcomes of prospective measurements on the constituents, even when they are far apart and noninteracting. However, as John Bell has taught to all of us, if the completeness assumption is made, it is illegitimate even to think that the constituents possess definite properties prior to the act of observation. In brief, the study of microphenomena has required radical conceptual changes and has led to a remarkable deepening in our understanding of nature.

When we come to the macroscopic domain, the superposition principle turns out to clash with [1] '*the definiteness, the particularity of the world*' as we experience it, and entanglement implies that, unless one limits 'in principle' (not simply FAPP) the class of observables for a system (a procedure which probably would require us [4, 5] to go beyond the Hilbert space formulation of quantum mechanics), even the apparata used in the measurement cannot be assumed to possess the desired properties specifying the outcomes. One could be tempted to consider this fact not as a fundamental difficulty, but simply as the indication that quantum mechanics is not universally valid and that, at the considered level (involving interactions of microscopic with macroscopic devices), it should be modified. For the time being, we should be satisfied with the fact that the formalism tells us how to deal with these situations by resorting to the postulate of wave packet reduction (WPR), which, as everybody knows, cannot be described by the linear laws of the theory. To those willing to take such an attitude, John Bell would have replied that, this being the situation, serious professionals should try to guess the more general laws of which those ruling microphenomena (the quantal ones) and those governing macroscopic systems (the classical ones) are simply good approximations. Apart from this, he would have stressed once more what he considered the most disturbing feature of the scheme: the theory does not contain any element which would consent to make a precise distinction between quantum and classical, the separation between these two classes of phenomena remaining basically shifty and, as such, allowing innumerable conceptually different but fundamentally vague ways out. All of them are characterized by the unacceptable (unless one is satisfied with FAPP attitudes toward science) shiftiness of the split: what is classical and what is quantum?, what is reversible and what is irreversible?, what is a measurement and what is an interaction (as such governed by the linear evolution) between two systems?, and finally, what is conscious [3]: '*a single celled living creature or a PhD student*'?

2.2. The dynamical reduction program (DRP)

As already stated, the 'beyond standard quantum mechanics' program we are going to describe corresponds to accepting a modification of the evolution law of the theory in such a way that all those measurement-like processes [3]: '*we are obliged to admit ... are going on more or less all the time, more or less everywhere*' have definite outcomes as a consequence of a unique dynamics. Such a dynamics should imply that the micro–macro interaction in a measurement process leads to WPR. With this in mind, it is appropriate to remark that the characteristic features distinguishing quantum evolution from WPR consist in the fact that, while Schrödinger's equation is linear and deterministic (at the wave function level), WPR is nonlinear and stochastic. It is then natural to consider the possibility of nonlinear and stochastic modifications of the standard hamiltonian dynamics.

Extremely interesting work in this direction had been done* and important results had been obtained [7] but, up to recent times, the program had not found a satisfactory formulation. Crucial problems had been left unsolved. In [8] John Bell wrote: '*The necessary theoretical development involves introducing what is called nonlinearity and perhaps what is called stochasticity into the basic Schrödinger equation. There have been interesting pioneer efforts in this direction, but not yet a breakthrough. This possible way ahead is unromantic in that it requires mathematical work by theoretical physicists, rather than interpretations by philosophers*'.

The first unsolved problem was the one of the choice of the preferred basis: if one is keen to consider the idea of a universal mechanism leading to reductions, within which linear manifolds should the reduction mechanism drive the state vector? The second open problem had been clearly identified by John Bell himself [9]: '*none of these modifications ... has the property required here of having little impact for small systems but nevertheless suppressing macroscopic superpositions. It would be good to know how this could be done*'. The problem of how it can happen that the reduction mechanism becomes more and more effective in going from the micro to the macro domain has been referred to [10] as the trigger problem. It represents the central point of the DRP.

2.3. Statistical ensembles and statistical operators, AND versus OR

There is a conceptual distinction which is extremely important to keep clearly in mind in order to fully appreciate the implications of the program. John Bell, in a letter addressed to A. Rimini, T. Weber and myself, after having expressed his interest in the papers, [11], in which for the first time a consistent and

* In particular we want to recall that [6] '*Philip Pearle kept alive this idea that it is some modification of the Schrödinger equation that is required*'.

satisfactory model (QMSL – see the next subsection) of dynamical reduction had been explicitly presented, felt the necessity of pointing out that, in his opinion, we had not been sufficiently clear on the following point: '... *in the reduced density matrix* ρ_Q *there are no off-diagonal elements. But this does not dispose of the embarrassing coherence of the superposition. On the other hand, your process of spontaneous reduction disposes of it very quickly*'. This pertinent remark focuses on the fundamental fact that since, as is well known, the correspondence between statistical ensembles (specified by their compositions in terms of pure subensembles) and statistical operators is, in general, infinitely-many to one, one could very well consider modifications of the dynamics leading to the diagonalization of the statistical operator in the appropriate 'preferred basis', which, however, do not suppress the unwanted superpositions of 'preferred states'. What has just been said is equivalent to the statement that to meet the macro-objectivistic requirements posed on it, the theory should guarantee that (under appropriate circumstances) individual reductions take place. As a consequence, the nice feature of QMSL of actually inducing such reductions could be better appreciated in a formulation directly using the state vector language than in one which, even though explicitly stated to deal with individuals, mainly makes use (as was the case for ref. [11]) of the statistical operator language.

The necessity of always making a clear distinction between statements about individuals and about ensembles has been repeatedly stressed, in different contexts, by John Bell. In particular, in his criticisms of the 'unproblematic formulations' or the 'easy solutions' of the measurement problem he pointed out that the replacement, even at the ensemble level, of the AND associated with the sum, with the OR referring to mutually exclusive properties possessed by the elements of appropriate subensembles of the physical ensemble one is dealing with, is illegitimate and derives from a 'probabilistic prejudice' inherited from the orthodox interpretation. In ref. [3], with reference to the expression:

$$\rho = \sum_n |c_n|^2 \Psi_n \Psi_n^* \tag{2.3}$$

he stated: '*if one were not actually on the lookout for probabilities, I think the obvious interpretation of even* ρ *would be that the system is in a state in which the various* Ψ*s somehow coexist:* $\Psi_1\Psi_1^*$ *and* $\Psi_2\Psi_2^*$ *and ... This is not at all a probability interpretation, in which the different terms are seen not as coexisting, but as* *alternatives:* $\Psi_1\Psi_1^*$ *or* $\Psi_2\Psi_2^*$ *or... The idea that the elimination of coherence, in one way or another, implies the replacement of "and" by "or" is a very common one among solvers of the measurement problem. It has always puzzled me.'*

To better focus the whole problem, let $|\Phi 1\rangle$ and $|\Phi 2\rangle$ be two states referring to two macroscopically different situations; one could, e.g., consider them to be the final states generated by triggering, with a spin 1/2 particle in an eigenstate of σ_z, an apparatus devised to detect the value of σ_z itself, so that:

$$|\Phi 1\rangle = |\text{spin up, pointer here}\rangle; \qquad |\Phi 2\rangle = |\text{spin down, pointer there}\rangle. \quad (2.4a)$$

Let us now consider a statistical ensemble E which is the union, with equal weights, of two subensembles E_i $(i = 1, 2)$, which are pure cases associated to the states $|\Phi i\rangle$, and another ensemble $\tilde{E}$ which is similarly composed of pure subensembles associated to the states $|A+\rangle, |A-\rangle$, where:

$$|A+\rangle = (1/\sqrt{2})[|\Phi 1\rangle + |\Phi 2\rangle], \qquad |A-\rangle = (1/\sqrt{2})[|\Phi 1\rangle - |\Phi 2\rangle]. \qquad (2.4b)$$

The two ensembles E and $\tilde{E}$ are manifestly described by the same statistical operator

$$\begin{aligned} \rho &= (1/2)|\Phi 1\rangle\langle\Phi 1| + (1/2)|\Phi 2\rangle\langle\Phi 2| \\ &\equiv (1/2)|A+\rangle\langle A+| + (1/2)|A-\rangle\langle A-|, \end{aligned} \qquad (2.5)$$

but, while for E it is perfectly legitimate to state that the pointer points either here or there since any member of the ensemble is described by one of the state vectors $|\Phi i\rangle$, each individual element of $\tilde{E}$ is still associated to a state vector exhibiting macroscopic coherence. Many of the proposed solutions to the measurement problem are based on the assumption that, for various reasons, the statistical operator at the end of the measurement of σ_z on a system which initially is in an eigenstate of σ_x can be assumed to have the form (2.5). This obviously cannot be rigorously but only FAPP correct. At any rate, even if one accepts this approximation one still has not overcome the difficulty. If one wants to be allowed to use the term OR, as required by the desire to account for the definiteness of the macroscopic record of the apparatus, one has to postulate that the final ensemble is precisely the one involving the states $|\Phi i\rangle$ and not $\tilde{E}$. But this means postulating that, during the process, WPR has occurred for the individual members of the ensemble.

We have discussed this point to stress that, when one is looking for a fundamental solution of the problem and, with this in mind, is keen to entertain the idea of describing WPR dynamically, one must guarantee that the evolution leads the state vector of each individual member of the ensemble within the appropriate manifolds (in the above example those spanned by $|\Phi i\rangle$). All the above matters have been reconsidered in great detail in ref. [12]. A precise analysis of the problem within the context of the DRP and a comparison with model theories which give only ensemble and not individual reductions can be found in refs. [13–15].

2.4. Quantum Mechanics with Spontaneous Localizations (QMSL)

Within the model [11, 16] which will be referred to as QMSL the problem of the choice of the preferred basis is solved by remarking that the most embarrassing superpositions, at the macroscopic level, are those involving different spatial locations of macroscopic objects. Actually, as Einstein has stressed [17], this is a crucial point which has to be faced by anybody aiming to take a macro-objective position about natural phenomena: '*A macro-body must always have a quasi-sharply defined position in the objective description of reality*'. Accordingly, QMSL considers the possibility of spontaneous processes, which are assumed to occur instantaneously and at the microscopic level, striving to suppress the linear superpositions of differently localized states. The required trigger mechanism must then follow consistently. The key assumption of QMSL is thus the following: each elementary constituent* of any system is subjected, at random times, to random and spontaneous localization processes (which we will call hittings) around appropriate positions. To have a precise mathematical model one has to be very specific about the above assumptions; in particular one has to make explicit HOW the process works, i.e. which modifications of the wave function are induced by the localizations, WHERE they occur, i.e. what determines the occurrence of a localization at a certain position rather than at another one, and finally WHEN, i.e. at what times, they occur. The answers to these questions are as follows.

Let us consider a system of N distinguishable particles and let us denote by $\Psi(\mathbf{q}_1, \mathbf{q}_2, \ldots, \mathbf{q}_N)$ the coordinate representation of the state vector (we

* John Bell has remarked [6] that, within QMSL, one should avoid the use of the expressions constituent or particle and should replace them by the expression 'argument of the wave function', since '*there are no particles in this theory, there is nothing but the wave function which has a certain number of arguments*'. In this chapter we will go on using the standard terminology.

disregard spin variables since the hittings are assumed not to act on them). The answer to the question HOW is then: if a hitting occurs for the i-th particle at point $\mathbf{x}$, the state vector $\Psi(\mathbf{q}_1, \mathbf{q}_2, \ldots, \mathbf{q}_N)$ changes instantaneously according to:

$$\Psi(\mathbf{q}_1, \mathbf{q}_2, \ldots, \mathbf{q}_N) \Rightarrow \Psi_{\mathbf{x}}(\mathbf{q}_1, \mathbf{q}_2, \ldots, \mathbf{q}_N) = \frac{\Phi_{\mathbf{x}}(\mathbf{q}_1, \mathbf{q}_2, \ldots, \mathbf{q}_N)}{\|\Phi_{\mathbf{x}}(\mathbf{q}_1, \mathbf{q}_2, \ldots, \mathbf{q}_N)\|^2} \tag{2.6}$$

$$\Phi_{\mathbf{x}}(\mathbf{q}_1, \mathbf{q}_2, \ldots, \mathbf{q}_N) = \left[\frac{\alpha}{\pi}\right]^{3/4} \exp\left[-\frac{\alpha}{2}(\mathbf{q}_i - \mathbf{x})^2\right] \Psi(\mathbf{q}_1, \mathbf{q}_2, \ldots, \mathbf{q}_N). \tag{2.7}$$

For what concerns the specification of WHERE the localization occurs, it is assumed that the probability density $P(\mathbf{x})$ of its taking place at the point $\mathbf{x}$ is given by:

$$P(\mathbf{x}) = \|\Phi_{\mathbf{x}}(\mathbf{q}_1, \mathbf{q}_2, \ldots, \mathbf{q}_N)\|^2, \tag{2.8}$$

so that hittings occur with a higher probability at those places where, in the standard quantum description, there is a higher probability of finding the particle. Note that the above prescription introduces nonlinear and stochastic elements in the dynamics.

Finally, the question WHEN is answered by assuming that the hittings occur at randomly distributed times, according to a Poisson distribution, with mean frequency λ.

It is immediate to convince oneself that the hitting process leads, when it occurs, to the suppression of the linear superpositions of states in which the same particle is well localized at different positions separated by a distance greater than $2/\sqrt{\alpha}$. As a simple example we can consider a single particle in the state (2.1) when $|\mathbf{h} - \mathbf{t}| \gg 1/\sqrt{\alpha} > \delta$. Suppose that a localization occurs around $\mathbf{h}$; the state after the hitting is then appreciably different from zero only in a region around $\mathbf{h}$ itself. A completely analogous argument holds for the case in which the hitting takes place around $\mathbf{t}$. For what concerns points which are far both from $\mathbf{h}$ and $\mathbf{t}$, one easily sees that the probability density for such hittings turns out to be, according to the prescription (2.8), practically zero, and moreover that if one of them occurred it would leave the wavefunction of the system almost unchanged.

2.5. The trigger mechanism

To understand the way in which the spontaneous localization mechanism is enhanced by increasing the number of particles which are in far apart (with respect to $1/\sqrt{\alpha}$) spatial regions, one can consider, for simplicity, the superposition $|\Psi\rangle$, with equal weights, of two macroscopic pointer states $|H\rangle$ and $|T\rangle$ corresponding to the pointer pointing at two different positions H and T, respectively. Taking into account that the pointer is 'almost rigid' and contains a macroscopic number N of microscopic constituents, the state $|\Psi\rangle$ can be written, with obvious meaning of the symbols:

$$|\Psi\rangle = K[|1 \cong h\rangle|2 \cong h\rangle \ldots |N \cong h\rangle + |1 \cong t\rangle|2 \cong t\rangle \ldots |N \cong t\rangle], \qquad (2.9)$$

where h is near to H, and t is near to T. The states appearing in (2.9) have coordinate representations $\langle \mathbf{q}_i|i \cong h\rangle$ and $\langle \mathbf{q}_i|i \cong t\rangle$ of the type (2.2). It is now evident that if any of the particles suffers a hitting process, e.g. near the point $\mathbf{h}$, the prescription (2.7) practically leads to the suppression of the second term in (2.9). Thus any spontaneous localization of any of the constituents amounts to a localization of, e.g., the centre of mass of the pointer. The hitting frequency is therefore effectively amplified proportionally to the number of constituents.

We stress that we have developed our argument by making reference, for simplicity, to an almost rigid body, i.e. to one for which all particles are around H in one of the states and around T in the other. It should, however, be obvious that what really matters in amplifying the reductions is the number of particles which are in different positions in the two states appearing in the superposition.

2.6. Choice of the parameters

The above argument allows one to understand easily how it happens that one can choose the parameters of the model in such a way that quantum predictions for microscopic systems remain fully valid, while the embarrassing macroscopic superpositions in measurement-like situations are suppressed in very short times. Correspondingly, individual macroscopic objects acquire, as a consequence of the unified dynamics governing all physical processes, definite macroscopic properties. The choice which has been proposed in ref. [11] is:

$$\lambda \cong 10^{-16}\,\mathrm{s}^{-1}, \quad 1/\sqrt{\alpha} \cong 10^{-5}\,\mathrm{cm}. \qquad (2.10)$$

It follows that a microscopic system suffers a localization, on the average, every 10^8 years, while a macroscopic one, every 10^{-7} seconds. John Bell comments [1]: within QMSL '*the cat is not both dead and alive for more than a split second'*. Besides the extremely low frequency of the hittings for microscopic systems, also the fact that the localization width be [6] '*large compared with the dimensions of atoms where quantum mechanics is essential*' so that even when a localization occurs '*it does very little violence to the internal economy of an atom*' plays an important role in guaranteeing [13] that no violation of well tested quantum mechanical predictions is implied by the modified dynamics.

Some remarks turn out to be appropriate. First of all QMSL, being precisely formulated, allows us to locate precisely the [3] '*shifty split*' between micro–macro, reversible–irreversible, quantum–classical. The transition between the two types of 'regimes' is governed by the number of particles which are well localized at distances more than 10^{-5} cm apart in the two states whose coherence is going to be dynamically suppressed. Secondly, the model is, in principle, testable against quantum mechanics. Actually, an essential part of the program consists in proving that its predictions do not contradict any established fact about microsystems and macrosystems. Before discussing this point, however, it turns out to be necessary to overcome some of the limitations of the model.

2.7. Continuous Spontaneous Localizations (CSL)

The QMSL model presented in the previous subsections has a serious drawback; as it has been formulated it does not allow one to deal with systems containing identical constituents. In fact [1]: '*an immediate objection to the GRW spontaneous wavefunction collapse is that it does not respect the symmetry or anti-symmetry required for "identical particles". But this will be taken care ... in the relativistic context ... I do not see why that should not be possible, although novel renormalization problems may arise.*' A quite natural idea to overcome this difficulty, as suggested by John Bell in 1987 in a private conversation, would be that of relating the hitting process not to a definite particle but to the particle number density averaged over an appropriate volume. However, the attempt to directly incorporate this idea in the QMSL scheme would require the introduction of a new constant besides the two which already appear in the model.

A more satisfactory treatment of this problem (see, however, the remarks at the end of this subsection) arose from a combination of the ideas [7] of P. Pearle with the specific mechanism considered in the formulation of QMSL. Such an approach has led to quite an elegant formulation of dynamical reduction mechanisms [10] in which the discontinuous jumps which are characteristic of QMSL are replaced by a continuous stochastic evolution (a sort of Brownian motion for the state vector) in the Hilbert space. The general framework has been discussed in ref. [18] (see also ref. [19]).

The CSL model is based on the consideration of a linear and stochastic evolution equation

$$\mathrm{d}|\Psi_B(t)\rangle = \left[\left(-iH - \frac{\gamma}{2}\sum_i A_i^2\right)\mathrm{d}t + \sum_i A_i\,\mathrm{d}B_i\right]|\Psi_B(t)\rangle, \tag{2.11}$$

where H is the hamiltonian of the system, $\{B_i\}$ is a Wiener process characterized by the drift and variance

$$\langle\langle \mathrm{d}B_i\rangle\rangle = 0, \quad \langle\langle \mathrm{d}B_i\,\mathrm{d}B_j\rangle\rangle = \gamma\delta_{ij}\,\mathrm{d}t \tag{2.12}$$

and $\{A_i\}$ is a set of self-adjoint commuting operators. We remark that Eq. (2.11) does not preserve the norm of the state vector but preserves its stochastic average. At this point one introduces nonlinearity by an assumption which parallels the one made within QMSL for the probability density of the hitting positions. One states that the probability density for the stochastic process B_i is not the one (denoted by $P_{\mathrm{raw}}[B_i]$) implied by Eq. (2.12), but is given by $P_{\mathrm{cooked}}[B_i]$, which depends also on the state vector at time t according to:

$$P_{\mathrm{cooked}}[B_i] = P_{\mathrm{raw}}[B_i]\,\||\Psi_B(t)\rangle\|^2. \tag{2.13}$$

One also stipulates that the physics at time t is determined by the normalized state vector $|\Psi_B(t)\rangle_N = |\Psi_B(t)\rangle/\||\Psi_B(t)\rangle\|$, $|\Psi_B(t)\rangle$ being the evolved state, according to Eq. (2.11) and with the particular realization B of the stochastic processes which actually occurred up to t, of the initial state vector.

To get a first hint of how the model works, let us suppose that the operators A_i have purely discrete spectra and let us consider their spectral representations

$$A_i = \sum_\sigma a_{i\sigma}P_\sigma. \tag{2.14}$$

If one disregards the hamiltonian evolution and considers the stochastic variables (summing up to 1):

$$z_\sigma(t) = {}_N\langle\Psi_B(t)|P_\sigma|\Psi_B(t)\rangle_N, \tag{2.15}$$

it is easy to prove [18] that one and only one of these variables tends to one for large times, for any specific stochastic history. Moreover, the probability that a given $z_\sigma(+\infty)$ takes the value 1 is given by its value at $t = 0$. In other words, the nonhamiltonian stochastic terms in the dynamics induce individual reductions on the common eigenmanifolds of the commuting operators A_i, with the required probabilities.

We can now formulate explicitly the proposed final version of the CSL model. It is obtained from the previous formalism by replacing the discrete index and the corresponding sum over i with a continuous index $\mathbf{x}$ and an integral over the variable $\mathbf{x}$, and the operators A_i with the average number density operator at the space point $\mathbf{x}$:

$$N(\mathbf{x}) = \left[\frac{\alpha}{2\pi}\right]^{3/2} \int d\mathbf{q} \exp\left[-\frac{\alpha}{2}(\mathbf{q}-\mathbf{x})^2\right] a^\dagger(\mathbf{q})a(\mathbf{q}), \qquad (2.16)$$

$a^\dagger(\mathbf{q})$ and $a(\mathbf{q})$ being the creation and annihilation operators for a particle at position $\mathbf{q}$. The parameter γ of Eq. (2.12) is chosen in such a way that the model reduces to QMSL in the case of a single particle*.

Once it is guaranteed that the dynamics induces individual reductions to states with almost definite number density, to understand its physically relevant implications one can make use of the statistical operator formalism. It is easy to see that the statistical operator obeys the equation:

$$\frac{d\rho}{dt} = -i[H,\rho] + \gamma \int d\mathbf{x} \left[N(\mathbf{x})\rho N(\mathbf{x}) - \frac{1}{2}\{N^2(\mathbf{x}),\rho\}\right]. \qquad (2.17)$$

Before discussing the main features of (2.17) we would like to remark that, in a sense, the idea of dealing with the problem of identical particles by using an average number and a further constant within an essentially QMSL-like hitting scheme was correct. In fact it has been possible to prove [18, 20] that for any CSL dynamics there is a hitting dynamics which is, from a physical point of view, 'as close to it as one wants'.

2.8 A simplified version of CSL

With reference to Eq. (2.17) and with the aim of understanding its physical implications such as the rate of suppression of coherence, we make some

*Obviously, when particles of different types (electrons, protons, etc.) are involved, in Eq. (2.16) a sum over an index specifying the type of particles will appear and the creation and annihilation operators will also carry such an index.

simplifying assumptions. First, we disregard the hamiltonian evolution; second, we divide the whole space into cells of volume $(2\pi/\alpha)^{3/2}$. Let us denote by $|n_1, n_2, \ldots\rangle$ the state in which there are n_i particles in cell i. From (2.17) one has:

$$\langle n_1, n_2, \ldots |\rho(t)|\tilde{n}_1, \tilde{n}_2, \ldots\rangle = \exp\left[-\frac{\gamma}{2}\left(\frac{\alpha}{4\pi}\right)^{3/2} \sum_i (n_i - \tilde{n}_i)^2 t\right] \times \langle n_1, n_2, \ldots |\rho(0)|\tilde{n}_1, \tilde{n}_2, \ldots\rangle. \quad (2.18)$$

The rate of suppression of the coherence of the two states $|n_1, n_2, \ldots\rangle$ and $|\tilde{n}_1, \tilde{n}_2, \ldots\rangle$ is thus determined by the expression

$$\exp\left[-\frac{\gamma}{2}\left(\frac{\alpha}{4\pi}\right)^{3/2} \sum_i (n_i - \tilde{n}_i)^2 t\right]. \quad (2.19)$$

On the basis of Eq. (2.19) one can study various physical effects. Apart from the differences which are related to the identity of the constituents, the overall physics is quite similar to the one implied by QMSL. Obviously, there are many interesting physical implications which deserve to be discussed. A detailed analysis has been presented in ref. [13]. As shown there and as follows from interesting estimates about possible effects in superconducting devices [21] and about excitations of atoms [22], it turns out not to be possible, with the present technology, to perform clear cut experiments allowing a discrimination of the model from standard quantum mechanics.

2.9. Motivations and achievements of the dynamical reduction program

We will now present a first summary about the previously described 'beyond conventional quantum mechanics' program by using John Bell's words.

Motivations [23]:
'One wants to be able to take a realistic view of the world, to talk about the world as if it is really there, even when it is not being observed... Our business is to try to make models and to see how far we can go with them in accounting for the real world'.

Achievements:

- Concerning the conceptual structure [1]: '*There is nothing in this theory but the wave function*'.

- Unifying properties [1]: '*In the GRW theory everything, including "measurement" goes according to the mathematical equations of the theory*'.
- Exactness [1]: '*Those equations are not disregarded from time to time on the basis of supplementary, imprecise, verbal, prescriptions*'.
- Macroscopic ambiguities [1]: '*Any embarrassing macroscopic ambiguity in the usual theory is only momentary in the GRW theory*'.
- Implications about locality [1]: '*The GRW theory does not add variables. But by adding mathematical precision to the jumps in the wave function, it simply makes precise the action at a distance of ordinary quantum mechanics*'.
- Limitations and promises [6]: '*The theories that I presented to you are certainly not beautiful. I think they are not true either; it may be they give some hint of where truth is to be found... I do think however that they have a certain kind of goodness – in the sense that they are honest attempts to replace the woolly words by real mathematical equations – equations which you don't have to talk away – equations which you simply calculate with and take the results seriously.*'

  And also [1]: '*I think that Schrödinger could hardly have found very compelling the GRW theory as expounded here – with the arbitrariness of the jump function and the elusiveness of the new physical constants. But he might have seen in it a hint of something good to come ... For myself, I see the GRW model as a very nice illustration of how quantum mechanics, to become rational, requires only a change which is very small (on some measures!).*'
- Relativistic requirements: The theory [1]: '*is as Lorentz invariant as it could be in the nonrelativistic version. It takes away the grounds of my fear that any exact formulation of quantum mechanics must conflict with fundamental Lorentz invariance.*'

The achievements of the DRP which are relevant for the debate about the foundations of quantum mechanics could also be concisely summarized by using the words of H.P. Stapp [24]: '*The collapse mechanisms so far proposed could, on the one hand, be viewed as ad hoc mutilations designed to force ontology to kneel to prejudice. On the other hand, these proposals show that one can certainly erect a coherent quantum ontology that generally conforms to ordinary ideas at the macroscopic level.'*

2.10. The 'when nobody looks' problem

The implications of the adoption of the Dynamical Reduction point of view can also be appropriately focused by considering the debate about properties possessed by individual physical systems. This debate has been nicely summarized in its essential aspects by D. Mermin in his brilliant presentation [25] of Bell's inequality. Mermin starts by quoting a recollection by A. Pais about Einstein:

'*We often discussed his notions on objective reality. I recall that during one walk Einstein suddenly stopped, turned to me and asked whether I really believed that the moon exists only when I took at it.'*

Then Mermin quotes Pauli:

'*... one should no more rack one's brain about the problem of whether something one cannot know anything about exists all the same, than about the ancient question of how many angels are able to sit on the point of a needle. But it seems to me that Einstein's questions are ultimately always of this kind.'*

D. Mermin is presenting Bell's inequality. He appropriately outlines how John Bell's work has radically changed the terms of the debate:

'Pauli and Einstein were both wrong. The questions with which Einstein attacked the quantum theory do have answers; but they are not the answers Einstein expected them to have. We now know that the moon is demonstrably not there when nobody looks.'

It is of some interest to see what conclusions one is led to draw about the above matter if one adheres to the DRP. The position should be obvious. Concerning Mermin's assertions, one would completely agree on the first two statements. One would also agree on the third, provided it were modified to refer to a microscopic system. So, one would agree that 'We now know that an electron in an EPR-Bohm like situation does not have the spin along any direction and also that an electron in a state like (2.1) is not at a definite place when nobody looks'. But, according to CSL, the statement is incorrect when referred to the moon: the moon is definitely there even if no sentient being has ever looked at it.

Since we have been led to mention Einstein's views about quantum mechanics we would like to conclude this part on nonrelativistic dynamical reduction models by raising the question of what Einstein's position about them could have been. If one takes into account what John Bell says about what Schrödinger's position could have been [1]: '*he might have seen in them a hint of something good to come*', mainly because '*he would not have been at all disturbed by their indeterminism. For ... he was expecting the fundamental laws to be statistical in character*', one would be led to conclude that Einstein would never have considered them seriously.

However, with reference to this remark, we would like to call attention to Einstein's fundamental work [26] of 1949. He considers first of all the problem of whether, within the quantum framework, it is reasonable to posit the existence of a definite time of decay of an unstable system. Then he takes into account the probable answer of the orthodox quantum theorist: he will point out that a determination of the disintegration time is not possible on an isolated system. Here Einstein counterargues by making resort to Schrödinger's remark that one meets the same difficulty when one considers the means for ascertaining the radioactive transformation. He assumes that the record consists of a mark on a strip of paper. Then he states: '*The location of the mark ... is a fact which belongs entirely within the sphere of macroscopic concepts ...*' and, as a consequence, '*there is hardly likely to be anyone who would be inclined to consider seriously ... that the existence of the location is essentially dependent upon the carrying out of an observation made on the registration strip. For, in the macroscopic sphere it simply is considered certain that one must adhere to the program of a realistic description in space and time; whereas in the sphere of microscopic situations one is more readily inclined to give up, or at least to modify, this program.*'

Some pages after he comes back to the same problem, and actually deepens his analysis by pointing out that '*the "real" in physics is to be taken as a type of program, to which we are, however, not forced to cling a priori.*' He adds: '*No one is likely to be inclined to attempt to give up this program within the realm of the "macroscopic" (location of the mark on the paper strip "real").*' And he concludes, '*But the "macroscopic" and the "microscopic" are so inter-related that it appears impracticable to give up this program in the "microscopic" alone.*'

We have felt the necessity of reporting such a long part of the text to point out that, for Einstein, the pursuit of realism was more a program which had been very successful than an a priori option, that he certainly refused FAPP solutions to the problems raised by the formalism (the mark on the paper strip must be described by the equations of the theory), but also that he would not have considered unacceptable, in principle, attempts giving up or wakening microrealistic requirements provided they would consent to take a macro-objective position. The dynamical reduction models achieve exactly this. Considering the above quoted sentences, one could be tempted to say that, in the case of Einstein, the main reasons* for not even contemplating such a way out seem to derive from the fact that it appears '*impracticable to give up this program in the microscopic alone*'. This position reflects the difficulty of imagining that the trigger problem could find a satisfactory solution since, as A. Shimony has stated when contemplating the possibility of dynamical reduction mechanisms [27]: '*Reasonable desiderata for such a theory pull in opposing directions*'. The analysis carried on so far has proved that, at least in the non-relativistic version, such a program is certainly practicable.

3. Relativistic dynamical reduction models

When confronted with a new theoretical scheme, particularly with one which, as we have seen [1] '*makes precise the action at a distance of ordinary quantum mechanics*', one is naturally led to raise the question of whether it is reasonable to think that it constitutes an approximation of a relativistically invariant theory. John Bell, who has always been deeply worried by nonlocality [28]: '*For me this is the real problem with quantum theory: the apparently essential conflict between any sharp formulation and fundamental relativity. That is to say, we have an apparent incompatibility,, at the deepest level, between the two fundamental pillars of contemporary theory.... It may be that a real synthesis of quantum and relativity theories requires not just technical developments but radical conceptual renewal*', has raised immediately this question for QMSL. Actually, as he himself has declared [6]: '*when I saw this theory first, I thought that I could blow it out of the water, by showing that it was grossly in violation of Lorentz invariance.*' The reason for this conviction is very simple: '*That's connected*

* We do not want to be misunderstood. We are not claiming that Einstein would easily have accepted the stochastic features of CSL. But he would certainly have appreciated the fact that, within such a model, the location of a macro-object does not depend on its being subjected to an observation and that, in the macroscopic sphere, the model does not require us to give up the program of a realistic description in space and time. Moreover, it might be useful to remark that, as Pauli pointed out in a letter [17] of 1954 to Born: '*Einstein does not consider the concept of "determinism" to be as fundamental as it is frequently held to be (as he told me emphatically many times)... he disputes that he uses as a criterion for the admissibility of a theory the question: "Is it rigorously deterministic?"*'.

with the problem of quantum entanglement, the Einstein–Rosen–Podolsky paradox.... The funny spooky action at a distance which Einstein saw in the ordinary formulation of quantum mechanics is not being removed by this reformulation, but emphasized. In the ordinary formulation you have a way out: maybe the wave function is not real – maybe there are things which are real which you never describe – maybe the things which are real are not behaving in a funny way. But in this new theory, there is nothing else but the wave function and the wave function is behaving in a funny way. It looks as if this could not possibly be Lorentz invariant, because of the long-range instantaneity.'

Thus, he started to investigate this point, to try to blow QMSL out of the water. After having performed a very detailed analysis [1, 6] by making resort to Dirac's multiple time formulation, he was led to the conclusion we have already mentioned in Subsection 2.9, i.e. that [1]: '*The theory is as Lorentz invariant as it could be in the nonrelativistic version*', thus taking away the reasons for his deep concern about the conflict of any exact formulation of quantum mechanics with relativity.

The fact that the theory does not immediately clash with relativistic requirements does not ensure, however, that it admits a fully relativistic generalization. Actually, at the Bruno Touschek lectures in Rome in 1987 John Bell concluded his talk by stating that '*Lorentz Invariance is now the big problem*' and in one of his last papers, after having expressed once more his dissatisfaction with the lack of precision of the standard formulation of quantum mechanics and the opinion that the only available acceptable alternatives are the Pilot Wave and QMSL, he stated [3]: '*The big question, in my opinion, is which, if either, of these two precise pictures can be redeveloped in a Lorentz invariant way*'.

Already in 1987 he had pointed out that in trying to follow this line [1] '*novel renormalization problems may arise*', and during the symposium in honour of J. Schwinger's seventieth birthday held at UCLA in March 1988, he raised the question of whether [6]: '*the jumping process could somehow be formulated within the Schwinger and Tomonaga and Stuckelberg way of presenting quantum field theory*' which seemed to him the most natural for this purpose. He also pointed out the '*immense difficulty*' one would meet in replacing the '*particle position*' of QMSL by a field variable, since '*all the local quantities in ordinary quantum field theory fluctuate infinitely*'. By trying to '*diminish this infinite fluctuation, you are going to create things in the vacuum*' and, as a consequence, contrary to what happens in the nonrelativistic case, '*you are going to do severe damage to the internal economy of the system*'. We will discuss below the appropriateness of these remarks. Before coming to this point it is useful to quote some interesting recent investigations about the non local aspects of CSL.

As is well known [29], Bell's locality assumption is equivalent to the conjunction of two other assumptions, viz., in Shimony's terminology [29], parameter independence and outcome independence*. In view of the experimental violation of Bell inequality, one has to give up either or both of these assumptions. The above splitting of the locality requirement into two logically independent conditions is particularly useful when discussing the different status of CSL and deterministic hidden variable theories with respect to relativistic requirements. In fact, as proved by Jarrett himself [29], when parameter independence is violated, if one had access to the variables which specify completely the state of individual physical system, one could send faster-than-light signals from one to the other wing of the apparatus. Even more, in ref. [30] it has been shown that it is not possible to build up a genuinely relativistic invariant# theory which, in its nonrelativistic limit, exhibits parameter dependence effects: any such theory entails backward causation. On the contrary, if locality is violated only by the occurrence of outcome dependence then faster-than-light signalling cannot be achieved since the stochastic outcome at a wing cannot be controlled.

Now it is well known that in a deterministic theory (i.e. one in which the range of any probability distribution of outcomes is the set $\{0, 1\}$) reproducing

*We recall that the first condition, parameter independence, is the requirement that the probability of getting an outcome at one wing of an apparatus in an EPR-like set-up is independent from the setting chosen at the other wing, while outcome independence amounts to requiring that the probability of an outcome at one wing does not depend on the outcome which is obtained at the other wing.

By the expression 'genuinely invariant' we mean, following Bell once more, a theory which does not admit even a hidden preferred reference frame [6]: (in hidden variable models of quantum field theories) '*you have something like the situation in relativity theory before Einstein, where there was a preferred reference frame – there was a significance in absolute simultaneity – but the Fitzgerald contraction and the Lorentz dilation fooled moving observers into thinking that light had the same velocity relative to them – so that they could even imagine themselves to be at rest. Now for me, this is an incredible position to take – I think it is quite logically consistent, but when one sees the power of the hypothesis of Lorentz invariance in modern physics, I think you just can't believe in it.... So that* (the existence of a preferred reference frame) *for me is a big defect. However, what I want to insist is that this theory agrees with experiment – when you dismiss this theory on the grounds of non-Lorentz invariance, you are requiring more than agreement with experiment. And I think it's very reasonable to require more than the agreement with experiment, I think theoretical physics owes much to insisting on more than agreement with experiment.*'

quantum predictions it is impossible to have outcome dependence, so that it must exhibit parameter dependence. This fact by itself suggest that such theories will certainly meet more serious difficulties with relativity than theories like standard quantum mechanics which exhibits only outcome dependence and does not allow faster-than-light signalling [31]. What about CSL? Recently it has been possible to prove [32] that it, too, violates Bell's locality only by violating outcome independence. This is to some extent encouraging; even though, as we will see in what follows, it seems extremely difficult to build a relativistic model inducing reductions, the now quoted result shows that there are no reasons of principle making such a project unviable.

3.1. Generalizing CSL

The first attempt to obtain a relativistic generalization of dynamical reduction models has been presented in ref. [33]. It has to be stressed that the requirement to have individual reductions forbids the theory to be invariant at the individual level (note that even QMSL and CSL are not Galilei invariant at the individual level). Thus one is led to introduce a generalization of the invariance requirement: the theory must be stochastically invariant. This means that, even though the individual processes may look different to different observers, any two of them will agree on the composition of the final ensemble for (subjectively) the same initial conditions. We remark that it is precisely in this sense that both QMSL and CSL turn out to be Galilei invariant.

In ref. [33] P. Pearle has considered a fermion field coupled to a meson field and has put forward the idea of inducing localizations for the fermions through their coupling to the mesons and a stochastic dynamical reduction mechanism acting on the meson variables. He considered Heisenberg evolution equations for the coupled fields and a Tomonaga–Schwinger CSL-type evolution equation with a skew-hermitian coupling to a *c*-number stochastic potential for the state vector.

This approach has been systematically investigated in refs. [15] and [34]. The general formalism can be briefly sketched as follows. One considers the Lagrangian density $L_0(x)$ describing the free evolution and hermitian couplings among the fields, and assumes that the fields evolve according to Heisenberg equations corresponding to $L_0(x)$. One also considers a stochastic

Tomonaga–Schwinger evolution equation for the state vector with a skew-hermitian coupling between a Lagrangian density $L_1(x)$ and a stochastic c-number potential $V(x)$:

$$\frac{\delta|\Psi_V(\sigma)\rangle}{\delta\sigma} = [L_1(x)V(x) - \lambda L_1^2(x)]|\Psi_V(\sigma)\rangle. \tag{3.1}$$

The densities $L_0(x)$ and $L_1(x)$ are Lorentz scalar functions of the fields. It is assumed that $L_1(x)$ does not contain derivatives of the fields. The c-number scalar function $V(x)$ is a white noise with covariance:

$$\langle\langle V(x)V(x')\rangle\rangle = \lambda\delta(x - x'). \tag{3.2}$$

Within the theory, the Tomonaga–Schwinger equation (3.1) plays the same role as the 'raw' equation (2.11) of CSL. Correspondingly, the evolved state has to be normalized and the potential requires a cooking procedure analogous to (2.13):

$$P_{\text{cooked}}[V(x)] = P_{\text{raw}}[V(x)]\||\Psi_V(\sigma)\rangle\|^2. \tag{3.3}$$

Disregarding the hamiltonian evolution, one can prove that the model yields individual reductions on the common eigenmanifolds of the commuting operators $L_1(x), x \in \sigma_0, \sigma_0$ being the space-like surface defining the initial conditions.

When the choices of ref. [33] are made for $L_0(x)$ and $L_1(x)$ and when some rough approximations are done, one gets, in the nonrelativistic limit, a CSL-type equation inducing space localizations. However, due to the white noise nature of the stochastic potential [1] *novel renormalization problems arise*: the increase per unit time and per unit volume of the energy of the meson field is infinite due to the fact that infinitely many mesons are created. As predicted by John Bell [6], in the search for a consistent relativistic dynamical reduction model one meets an *immense difficulty*, one ends up creating *things in the vacuum* and doing *severe damage to the internal economy of the system.*

3.2. Achievements of the relativistic dynamical reduction program

For the reasons we have just discussed, one cannot say that the possibility of generalizing CSL to the relativistic case has been proved. Even some recent trials in this direction [35] did not allow us to completely overcome the difficulties. In spite of this we think that the efforts which have been spent on such a program have led to a better understanding of some points and have thrown some light on important conceptual issues. As J. Bell has stated in his lecture at Trieste in November 1989, on the occasion of the celebrations of the 25th anniversary of the establishment of the International Centre for Theoretical Physics: people involved in the pursuit of the relativistic program *have met some interesting difficulties.*

We consider it useful to list some of the achievements of this line of research. First, it has led to a completely general and rigorous formulation [15, 34, 35] of the concept of stochastic invariance. Second, it has stimulated a critical revisitation, based on the consideration of smeared observables with compact support, of the problem of locality at the individual level. This analysis has put into evidence the necessity of reconsidering the criteria for the attribution of objective local properties to physical systems. A way to do this has been proposed, with the following implications. In specific situations one cannot attribute any local property to a microsystem; any attempt to do so gives rise to ambiguities. However, in the case of macroscopic systems, the impossibility of attributing to them local properties (or, equivalently, the ambiguities about such properties) lasts only for time intervals of the order of those which are necessary for the dynamical reduction to take place. Moreover, no objective property corresponding to a local observable can emerge, even for microsystems, as a consequence of a measurement-like event occurring in a space-like separated region. Such properties emerge only in the future light cone of the considered macroscopic event. Finally, an analysis based on counterfactual arguments [30, 34, 36] has shown that the very formal structure of the theory is such that it does not consent, even conceptually, to establish cause–effect relations between space-like events. In some sense one could state that even in the relativistic version, the DRP [8] '*allows electrons* (in general microsystems) *to enjoy the cloudiness of waves, while allowing tables and chairs, and ourselves, and black marks on photographs, to be rather definitely in one place rather than another, and to be described in classical terms*'.

After having listed some interesting results obtained along this line, in concluding this section we feel the necessity to stress once more the immense difficulties that the program of a relativistic generalization has met up to now. The question of whether such a program will find a satisfactory formulation still remains '*the big problem*' for this type of investigation.

4. Reply to criticisms and hints

The DRP in general, and more specifically QMSL, have been the object of various criticisms. In this section we will take into account those which seem more relevant to us.

4.1. The criticism of 'being ad-hoc'

To start with we will mention a quite general objection against the DRP which has been raised by various authors in different contexts. It consists in the assertion that, since no attempt is made at introducing a fundamental physical mechanism accounting for the occurrence of the spontaneous localizations, the model turns out to be very ad-hoc.

Concerning this criticism we must simply say that, in principle, we agree: up to now, no specific proposal for the identification of a physical mechanism giving rise to the stochasticity has been presented.

There are, however, at least two partial answers to this criticism which deserve to be presented. The first one has been given by John Bell himself [6] ... '*it would be good to explain them* (the spontaneous reductions)*; but you can't explain everything, you have to start somewhere. It would be good to relate these quantum jumps which I am now introducing to some big idea and not just to bring them out of a hat like that – but for the moment that's just what we have to do. Just as in the early days of radioactivity theory we had to say that our radioactive nucleus just at some moment goes pop! – it decays – for reasons unknown*'. It is interesting to note that in the same paper J. Bell pointed out the goodness and honesty of these precise attempts.

Secondly, we would like to stress that the correct way of judging such attempts requires us to look at them within the appropriate context, i.e. the one of the debate about the foundations of quantum mechanics. If one chooses this perspective one can compare, e.g., the achievements of the DRP with those of the hidden variable models. With reference to the most satisfactory available model of this type, i.e. the de Broglie–Bohm Pilot Wave theory, one could also raise the objection that it is ad hoc in its assumption that specifically the position* variables are the extra variables to be added to the wave function

*In fact it is not difficult to devise deterministic hidden variable models in which the preferred variables are different, e.g., they could be the momenta of the particles.

for a complete and precise description of physical systems. Analogously, one could consider ad hoc the [6] '*one more hypothesis*' one needs '*to connect this scheme with ordinary nonrelativistic quantum mechanics ... God chose first the initial wave function* $\Psi(0)$, *and then the initial configuration of the particles* $\mathbf{x}(0)$. *He chose this configuration at random or she chose it at random, from an ensemble in which the distribution was given at the initial time by the square of the wave function*'. In spite of these features* of the model one could ask: is it not of remarkable conceptual relevance to know that, on one side, a deterministic completion of quantum mechanics is actually possible, and, on the other, that the price to pay to adopt such a point of view is to accept contextuality? In complete analogy one could raise the question: is it not conceptually relevant to have the explicit proof that there are theories [3] '*which have nothing in their kinematics but the wave function*' and nevertheless allow us to take a macro-objective realist position about natural phenomena? Is it not relevant to have learned that the price to pay# consists in accepting both stochastic and non-linear modifications of the evolution equation?

Concerning the alleged arbitrariness of the two parameters appearing in the considered models (α and λ in QMSL, γ and λ in CSL), it is worth remarking that, if one takes into account necessary physical requirements, they turn out not to be, after all, so arbitrary. In fact, the localization parameter must be appreciably larger than atomic dimensions [6] *to avoid doing violence to the internal economy of an atom*, but relatively small on a macroscopic scale. Furthermore, the allowed range of variability of the product $\alpha\lambda$ turns out to be rather narrow [13] if one wants to be sure that the model does not entail unacceptable consequences concerning, e.g., the mean energy increase or the dissociation of atoms. It has also been proved [39] that the detailed shape of the jump function is to a large extent irrelevant and that [40] reduction mechanisms involving other variables (e.g. momenta) beside or in place of the positions (which play such a privileged role within QMSL), cannot be considered as acceptable candidates for the DRP: they are either ineffective at

* Actually, in recent times, extremely important investigations [37] leading to a remarkable clarification of this delicate point have been presented.

In the study of dynamical reduction models it has been possible to prove [19] that non-linearity without stochasticity leads to faster than light effects, while stochasticity without nonlinearity is unable to yield [38] individual reductions and allows us only to describe ensemble reductions.

inducing definite properties for individual macroscopic objects, or give rise to unacceptable consequences for the microscopic ones.

To conclude, we consider it appropriate to comment on the very appearance of the (ad hoc) parameters which would play, if one took the models of dynamical reduction proposed so far very seriously, the role of new constants of nature. The DRP is primarily concerned in describing how, and in making precise where and when, the transition from quantum to classical, from almost exactly reversible to almost definitively irreversible, takes place. It has to be stressed that the parameter 'the number of particles which are involved in the physical process one is dealing with' cannot be assumed to characterize, by itself, the transition regime. In fact, as we all know, there certainly are macroscopic systems which exhibit a genuine quantum behaviour and which require a full quantum treatment. Since, however, as repeatedly stressed, the main aim of the DRP is that of forbidding, in the case of a macroscopic object, the occurrence of linear superpositions of far-away states, the specification of the level at which the decoherence mechanism becomes effective must involve a scale for distinguishing here from there. Within QMSL and CSL such a scale is given by the localization parameter $1/\sqrt{\alpha}$. The above argument does not render unreasonable the idea that any theory describing in a mathematically precise way the desired loss of coherence requires the introduction of new constants*. Obviously this fact cannot be considered as a strict argument in favour of new parameters but it makes less objectionable [1] '*the elusiveness of the new physical constants*'.

We come now to other specific criticisms. They have been put forward in various papers by different authors; a quite exhaustive list of them has been presented by A. Shimony in a recent work [43]. The author agrees that the DRP is interesting and deserves great attention, but he lists eight desiderata that such a program should satisfy. Out of these, in his opinion, four are met by QMSL and CSL, the remaining ones are not. We will analyze them in what follows.

*Recently an attempt [41] to get rid of the two constants appearing in QMSL has been done. However this proposal has [42] unacceptable physical consequences. In ref. [42] a modification of the model of ref. [41] has been presented, resulting in a dynamical reduction theory involving only one parameter (specifying a length scale) and relating reduction to gravity.

4.2. Relativistic invariance

As required by Shimony: *the modified dynamics should be Lorentz invariant.* We perfectly agree; nevertheless, as it should be clear from the discussion of Section 3, we believe that this cannot be considered as a criticism to the presently available models of dynamical reduction, but it has rather to be taken as a recognition that hard work has still to be done by those who consider it interesting to try to fully exploit this line of thought.

4.3. Preclusion versus suppression of macroscopic coherence

According to the author of ref. [43]: *the modified dynamics should preclude the gestation of Schrödinger's cat, and in general the occurrence – even for a brief time – of states of a system in which a macroscopic variable is indefinite.* We must confess that we have not been able to grasp the meaning of this requirement. Since the theory pretends to describe the measurement process which involves the interaction of the measured system with the measuring apparatus, does this requirement pretend that the triggering of the apparatus and the stochastic choice of the outcome should occur instantaneously? How could that be?

Probably some clarification about this point can come from the following remark. The criticism seems to hypothesize the actual possibility of occurrence of a superposition of, e.g., states describing a macroscopic pointer pointing here and pointing there. We recall some debates with colleagues in which objections starting with specifications of the type: *if you consider a linear superposition with equal weights of this table being in Geneva and in Rome ... and a localization process ...* With reference to remarks of this kind it is extremely important to have clear that, if the dynamics characterizing QMSL is actually operative, it forbids the occurrence of such situations. What actually happens is the following. Suppose we trigger, e.g., a macrosystem with a linear superposition of two microstates. Assume further that the conventional hamiltonian dynamics describing the system apparatus interaction would, for one of the initial microstates, leave the table at rest here in Geneva, while for the other it would deliver to it a momentum pointing to Rome. Then, according to Eq. (2.19), as soon as 10^9–10^{10} particles are displaced more than 10^{-5} cm, the modified dynamics compels the table either to remain at rest or to move towards Rome. The specification either–or refers precisely to the suppression of one of the two quantum states. To try to bring this suppression to the precise instant (?) of the system–apparatus interaction seems to us physically meaningless.

4.4. Actual versus idealized measurements with definite outcomes

This criticism by Shimony admittedly makes reference to an objection repeatedly [44] raised against QMSL. It is based on the remark that one can easily imagine situations leading to definite perceptions and which nevertheless do not involve the displacement of a large number of particles up to the stage of the perception itself. These cases would then constitute *actual* measurement situations which cannot be described by QMSL, contrary to what happens for the *idealized* (!) situations considered by such a model, involving the displacement of some sort of pointer. To be more specific, in ref. [44] consideration has been given to a 'measurement' process whose output is the emission of a burst of photons. More precisely, one considers a Stern–Gerlach set-up in which the two paths followed by the microsystem according to the values of its spin component hit a fluorescent screen, exciting a small number of atoms which subsequently decay by emitting a small number of photons. The argument goes as follows: since only a few atoms are excited, since the excitations involve displacements which are smaller than the characteristic localization distance of QMSL, since QMSL does not induce reductions on photon states and finally since the photon states overlap immediately, there is no way for the spontaneous localization mechanism to become effective. The superposition of the states 'photons emerging from point A of the screen' and 'photons emerging from point B of the screen' will last for a long time. On the other hand, since the visual perception threshold is quite low (about 6–7 photons), there is no doubt that the naked eye of a human observer is sufficient to detect whether the luminous spot on the screen is at A or at B. Then, the authors of the above papers draw their conclusion: within QMSL no dynamical reduction can take place and as a consequence no measurement is over, no outcome is definite, up to the moment in which a conscious observer perceives the spot.

We have presented a detailed answer to this criticism in a recent paper [45]. The crucial points of our argument are the following: we perfectly agree that in the case considered the superposition persists for long times (actually the superposition must persist, since, the system under consideration being microscopic, one could perform interference experiments with it, which everybody would expect to confirm quantum mechanics). Thus, if one takes the above remark seriously, one is compelled to consider the actual systems which enter into play and the changes induced in them by the photon beams. We have made a simple estimate of the number of ions which are involved in the visual perception mechanism; such an analysis makes perfectly plausible that, in the process, a sufficient number of particles are displaced by a sufficient spatial amount to satisfy the conditions which are necessary, according to QMSL, for the suppression of the superposition of the two nervous signals to take place within the perception time.

We do not want to be misunderstood: this analysis does by no means amount to attributing a special role to the conscious observer or to his perception. The observer's brain is the only system present in the set up in which a superposition of two states involving different locations of a large number of particles occurs. As such it is the only place where the reduction can and actually must take place according to the theory. After all, in a serious theory [3] '*not just vague words but precise mathematics has to tell us what is system and what is apparatus, which natural processes have the special status of measurements*'. We consider it of extreme relevance to stress that if in place of the eye of a human being one puts in front of the photon beams a spark chamber or a device leading to the displacement of a macroscopic pointer or producing ink spots on a computer output, reduction will take place. In the considered example, the human nervous system is simply a physical system, a specific assembly of particles, which makes the same function as one of these devices, if no such device is interacting with the photons before the human observer does. There follows that it is incorrect and seriously misleading to claim that QMSL requires a conscious observer to make definite the macroscopic properties of physical systems.

A further specification is appropriate. The previous analysis could be taken by the reader as indicating that we adopt a very naive and oversimplified attitude about the deep problem of the brain–mind correspondence. We do not claim and we do not pretend that QMSL allows a physicalist explanation of conscious perception. We simply point out that, for what we know about the purely physical aspects of the process, we can state that before the nervous pulses reach the higher visual cortex, the conditions guaranteeing the suppression of one of the two signals are verified. In brief, a consistent use of the dynamical reduction mechanism in the above situation makes not unreasonable the definiteness of the conscious perception.

4.5. The problem of the tails of the wavefunction

The last desideratum for a dynamical reduction model which, according to some authors, is not met by QMSL is that [43] '*one should not tolerate tails in wave functions which are so broad that their different parts can be discriminated by the senses, even if very low probability amplitude is assigned to them.*' This is, in our opinion, the most interesting criticism since it allows us to focus on a deep problem, and to point out two interesting points of view by John Bell about it, as well as some recent proposals which represent, in our opinion, the appropriate interpretation of dynamical reduction models.

We start by considering the situation within the conventional quantum framework. The square of the modulus of the wave function in configuration space gives the position probability density of, e.g., a particle. We agree that the fact that wave functions never have strictly compact spatial support can be considered as puzzling. However, this is a problem arising directly from the mathematical features (spreading of wave functions) and from the probabilistic interpretation of the theory, and not at all a peculiar problem of the dynamical reduction models. For this purpose it is appropriate to remark that nobody considers the fact that, e.g., the wave function of a pointer or of a table does not have a compact support in the centre of mass coordinate as representing a difficulty for quantum mechanics. When the wave function is extremely well peaked around a fixed space point in the considered variable, one accepts that the wave function itself describes a table which is located at a certain position and that this corresponds in some way to our perception about it. It is obviously true that, for the considered wave function, the quantum rules entail that if a measurement were performed the table could be found (with an extremely small probability) to be kilometres away, but this is not the measurement or the so called macro-objectification problem of the standard theory. The real question at issue here is that of the possibility of attributing local objective properties to physical systems. Such a problem can find a solution only through an appropriate definition of the instances which render such an attribution legitimate. We have repeatedly discussed this point; see, e.g., [12, 14, 15 and 34].

Obviously, to close the circle and have a fully satisfactory world-picture, one should relate the possibility of attributing definite properties to physical systems to the conditions of human experience. This is the ancient problem which haunted Descartes and Locke. We do not want to take a precise position about it (this is a specific task for the philosophers), we simply invoke the necessity of accepting at least a vague correspondence between the only formal elements of the theory (the wave function) and our perceptions. Consequently, the above state of affairs can be reasonably taken as accounting for our perception that the table is at the considered place.

A completely different situation arises when one considers a superposition, with equal weights, of two wave functions, both possessing tails (i.e. having non-compact support) but both being appreciably different from zero only in very narrow intervals, which, however, are macroscopically separated. This is the really embarrassing situation which conventional quantum mechanics is unable to make understandable. To which perception about the position of the table does the considered wavefunction correspond?

The implications for this problem of the adoption of QMSL should be obvious. Within QMSL, the considered superposition of two states which, when considered individually, are assumed to correspond to different and definite perceptions about macroscopic locations, are dynamically forbidden. If some process strives to produce such superpositions, then the nonhamiltonian dynamics induces the localization of the centre of mass and consequently the possibility of attributing to the system the property of being in a definite place. Correspondingly, the possibility arises of accounting for our definite perception about it.

Coming back to the criticism, we remark that the requirement that the appearance of macroscopically extended (even though extremely small) tails be strictly forbidden is motivated by the choice of strictly committing oneself to the probabilistic interpretation of the theory even for what concerns the psycho-physical correspondence: states assigning non-exactly vanishing probabilities to different outcomes of position measurements must correspond to ambiguous perceptions about these positions. Since neither within the standard formalism nor in the framework of dynamical reduction models a wave function can have compact support, taking such a position leads us to conclude that it is just the Hilbert space description of physical systems which has to be given up.

These remarks lead us in a natural way to present the interesting and original ways suggested by John Bell to overcome the difficulties of the probabilistic interpretation within the QMSL and CSL descriptions of physical processes. The first suggestion appears in the first paper [1] he wrote about QMSL and is strictly related to the discrete structure of the model. In QMSL the spontaneous localizations of particles take place, as we have seen, at precise places and at precise time (even though governed by probabilistic laws). Thus one can take these space–time events as the fundamental quantities the theory is about. It is appropriate to present this view by resorting to John Bell's worlds: *There is nothing in this theory but the wave function. It is in the wave function that we must find an image of the physical world, and in particular of the arrangement of things in ordinary 3-dimensional space. But the wave function as a whole lives in a much bigger space, of 3N dimensions. It makes no sense to ask for the amplitude or phase or whatever of the wave function at a point in ordinary space. It has neither amplitude nor phase nor anything else until a multitude of points in ordinary 3-space are specified. However, the GRW jumps (which are part of the wave function, not something else) are well localized in ordinary space. Indeed each is centred on a particular space–time point* $(\mathbf{x}, t)$*. So we can propose these events as the basis of the 'local beables' of the theory. These are the mathematical counterparts in the theory to real events at definite places and times in the real world (as distinct from the many purely mathematical constructions that occur in the working out of physical theories, as distinct from things which may be real but not localized, and as distinct from the 'observables' of other formulations of quantum mechanics, for which we have no use here). A piece of matter then is a galaxy of such events. As a schematic psychophysical parallelism we can suppose that our personal experience is more or less directly of events in particular pieces of matter, our brains, which events are in turn correlated with events in our bodies as a whole, and they in turn with events in the outer world.*

Subsequently he considered also another possibility, which he repeatedly proposed in public talks and in private conversations and correspondence and which I am going to espouse. In 1989, at Erice, during a party, I was listening to him speaking with Pearle, Shimony and other people. The place was noisy and I could not follow the argument completely, but I grasped some statements addressed to P. Pearle of the type: *you are still too strictly committed to the probabilistic orthodox interpretation! You have to go beyond it! Now you are allowed to take a density interpretation!*. Early the following morning, I met

John Bell and I asked him whether he intended to suggest that, if Schrödinger had known and adopted the QMSL point of view, he would have not been compelled to abandon his density interpretation for the wave function and to accept the probabilistic interpretation of the Copenhagen school. John Bell's answer was sharp: *I meant precisely that!* He came back to this point on various occasions. In ref. [3] he wrote: '*The GRW-type theories have nothing in their kinematics but the wave function. It gives the density (in a multidimensional configuration space!) of stuff. To account for the narrowness of that stuff in macroscopic dimensions, the linear Schrödinger equation has to be modified, in the GRW picture by a mathematically prescribed spontaneous collapse mechanism*'. We have discussed this point repeatedly. I cannot avoid quoting the letter he sent to me on October 3, 1989: '*As regards Ψ and the density of stuff, I think it is important that this density is in the 3-N dimensional configuration space. So I have not thought of relating it to ordinary matter or charge density in 3-space. Even for one particle I think one would have problems with the latter. So I am inclined to the view you mention "as it is sufficient for an objective interpretation ..." And it has to be stressed that the "stuff" is in 3-N space – or whatever corresponds in field theory.*'

It may have been just the repeated exchanges of view that I had with John Bell on this matter which stimulated the elaboration of a different approach [46] which, in our opinion, represents the most appropriate way to 'close the circle' within the context of a genuinely Hilbert space description of natural processes, when the evolution law is of the QMSL or CSL type.

The analysis started with a remark about the inappropriateness of the concept of 'distance' in the Hilbert space to account for the similarity or difference of macroscopic situations. If one considers three states $|\varphi_H\rangle$, $|\varphi_H^*\rangle$ and $|\varphi_T\rangle$ of a macrosystem (let us say a massive macroscopic bulk of matter), the first corresponding to its being located here, the second to its having the same location but one of its atoms (or molecules) being in a state which is orthogonal to the corresponding state in $|\varphi_H\rangle$, and the third having exactly the same internal state of the first but being differently located (there), then, in spite of the fact that the two first states are 'practically indistinguishable from each other at the macrolevel' while the first and the third correspond to completely different and directly perceivable situations, the distance $\|(|\varphi_H\rangle - |\varphi_H^*\rangle)\|$ is equal to the one $\|(|\varphi_H\rangle - |\varphi_T\rangle)\|$.

The second step consisted in remarking that, in the most general case (i.e. even when one is dealing with a body which is not almost rigid, such as a gas or a cloud), the mechanisms leading to the suppression of the superpositions of macroscopically different states are governed fundamentally (recall Eq. (2.19)) by the sum (or the integral) of the squared differences of the mass densities associated to the two superposed states, averaged over the characteristic volume of the theory, i.e. $10^{-15}\,\mathrm{cm}^3$. This suggests that we should take the following attitude: what the theory is about, what is real 'out there' at a given space point $\mathbf{x}$ is just the average mass density in the characteristic volume around $\mathbf{x}$:

$$\mathfrak{M}(\mathbf{x}, t) = \langle \Psi(t)|M(\mathbf{x})|\Psi(t)\rangle, \tag{4.1}$$

where $M(\mathbf{x})$ is the mass density operator corresponding to the characteristic volume around $\mathbf{x}$ (i.e. it has a form analogous to (2.16) with the appropriate mass factors added). It is obvious that within standard quantum mechanics such a function cannot be endowed with any objective physical meaning, due to the occurrence of linear superpositions of macroscopically different mass distributions which give rise to functions $\mathfrak{M}(\mathbf{x}, t)$ which do not correspond to what we find in a measurement process or what we perceive (typically the equal weight superposition of $|\varphi_H\rangle$ and $|\varphi_T\rangle$ will give rise to a mass density distribution which is half of the actual one in both regions H and T). But in a CSL model relating reductions to mass density differences, the dynamics, as we have seen, suppresses the embarrassing superpositions in extremely short times. Then, if one considers only the states which are allowed one can give a description of the world in terms of $\mathfrak{M}(\mathbf{x}, t)$, i.e. one recovers a physically meaningful account of physical reality in the usual 3-dimensional space and time. Resorting to the quantity (4.1) one can also define an appropriate 'distance' between two states*:

$$\Delta(|\Phi\rangle, |\Psi\rangle) = \left\{ \int \mathrm{d}\mathbf{x}[\langle \Phi|M(\mathbf{x})|\Phi\rangle - \langle \Psi|M(\mathbf{x})|\Psi\rangle]^2 \right\}^{1/2}, \tag{4.2}$$

which turns out to be perfectly appropriate to ground the concept of macroscopically similar or distinguishable Hilbert space states. In turn (recall the remarks of Section 4.4) this distance can be used as a basis to define a sensible principle of psychophysical correspondence for the theory.

Needless to say, the presented interpretation now (like that proposed by Bell) gives a radical solution to the problem of the tails in the macroscopic case. Not

* Strictly speaking, such a quantity is not a distance, since it may happen that it vanishes even though the two states in its arguments do not coincide (it has, however, all other properties of a distance). But this fact, as the reader will easily understand, is not relevant for the following considerations.

only do the states $|\varphi_H\rangle$ and $|\varphi_H^*\rangle$ mentioned previously have (as they should) practically zero distance while the states $|\varphi_H\rangle$ and $|\varphi_T\rangle$ are 'very distant' from each other, but it also happens that the state $|\varphi_H\rangle$, and the state obtained from it by disregarding the exponential tail outside the region in which the wave function is appreciably different from zero, are very near to each other and can be identified for what concerns the macroscopic 'status' of the universe. One can then easily understand the appropriateness of the remark by John Bell referred to above: '*So I have not thought of relating it to ordinary matter or charge density in 3-space. Even for one particle I think one would have problems with the latter*'. The mass density functional $\mathfrak{M}(\mathbf{x}, t)$ cannot be endowed with any 'objective' meaning at the microlevel, just because states corresponding to different locations of microscopic systems can persist for extremely long times even within CSL. But, as repeatedly stressed in this paper, the very aim of the DRP is just [8] '*to allow electrons (in general microsystems) to enjoy the cloudiness of waves, while allowing tables and chairs, and ourselves, and black marks on photographs, to be rather definitely in one place rather than another, and to be described in classical terms*'.

These remarks conclude our exposition. We have hope of having been able to show that John Bell has played an important role in suggesting new ways and in stimulating attempts to give a satisfactory answer to the deeper questions raised by quantum mechanics. It has to be stressed that the possibility of getting such a result rests entirely on the adoption of the 'beyond conventional quantum mechanics' point of view outlined in this chapter.

ACKNOWLEDGMENTS

I am deeply grateful to Mary Bell, Daniele Amati and John Ellis for having offered me the opportunity of honouring John Bell, whose profound work and whose deep ideas have been a continuous source of inspiration for me and for all colleagues and friends involved in the research program discussed in this chapter.

REFERENCES

[1] J.S. Bell, in: *Schrödinger – Centenary Celebrations of a Polymath*, C.W. Kilmister, ed. (Cambridge University Press, Cambridge, 1987).

[2] K. Gottfried and N.D. Mermin, *Europhysics News* **22** (1991) 67.

[3] J.S. Bell, in: *Sixty-two Years of Uncertainty*, A. Miller, ed. (Plenum, New York, 1990).

[4] E.P. Wigner, in: *Foundations of Quantum Mechanics*, B. d'Espagnat, ed. (Academic Press, New York, 1971), 122–5.

[5] P. Busch, in: *Symposium on the Foundations of Modern Physics 1990*, P. Lahti and P. Mittelstaedt, eds. (World Scientific, Singapore, 1990). P. Busch, P.J. Lahti and P. Mittelstaedt, *The Quantum Theory of Measurement*, Lecture Notes in Physics (Springer-Verlag, 1996).

[6] J.S. Bell, in: *Themes in Contemporary Physics II, Essays in Honour of Julian Schwinger's 70th birthday*, S. Deser and R.J. Finkelstein, eds. (World Scientific, 1989).

[7] F. Karolyhazy, *Nuovo Cim.* **42A** (1966) 390; P. Pearle, *Phys. Rev.* **D134** (1976) 857; *Int. Jour. Theor. Phys.* **48** (1979) 489; *Found. Phys.* **12** (1982) 249; *Phys. Rev.* **D29** (1984) 235; N. Gisin, *Phys. Rev. Lett.* **52** (1984) 1657; *ibid.* **53** (1984) 1776.

[8] J.S. Bell, in: *Proceedings of the Nobel Symposium 65: Possible Worlds in Arts and Sciences* (Stockholm, 1986).

[9] J.S. Bell, in: *Quantum Gravity 2*, C. Isham, R. Penrose and D. Sciama, eds. (Clarendon Press, Oxford, 1981).

[10] P. Pearle, *Phys. Rev.* **A39** (1989) 2277.

[11] G.C. Ghirardi, A. Rimini and T. Weber, in: *Quantum Probability and Applications*, L. Accardi and W. von Waldenfels, eds. (Springer, Berlin 1985); *Phys. Rev.* **D34** (1986) 470; *ibid.* **D36** (1987) 3287.

[12] G.C. Ghirardi, in: *Quantum Chaos–Quantum Measurement*, P. Cvitanovic *et al.*, eds. (Kluwer Academic Publishers, Dordrecht, 1992).

[13] G.C. Ghirardi and A. Rimini, in: *Sixty-two Years of Uncertainty*, A. Miller, ed. (Plenum, New York, 1990).

[14] G.C. Ghirardi and P. Pearle, in: *PSA 1990*, **2**, A. Fine, M. Forbes and L. Wessels, eds. (Philosophy of Science Association, East Lansing, Michigan, 1990).

[15] G.C. Ghirardi, R. Grassi and P. Pearle, *Found. Phys.* **20** (1990) 1271.

[16] G.C. Ghirardi, A. Rimini and T. Weber, *Found. Phys.* **18** (1988) 1; F. Benatti, G.C. Ghirardi, A. Rimini and T. Weber, *Nuovo Cim.* **100B** (1987) 27.

[17] W. Pauli in: M. Born, *The Born Einstein letters* (Walter and Co., New York, 1971).

[18] G.C. Ghirardi, P. Pearle and A. Rimini, *Phys. Rev.* **A42** (1990) 78.

[19] N. Gisin, *Helv. Phys. Acta* **62** (1989) 363.

[20] O. Nicrosini and A. Rimini, *Found. Phys.* **20** (1990) 1317.

[21] A.I.M. Rae, *J. Phys.* **A23** (1990) 57; M.R. Gallis and G.N. Fleming, *Phys. Rev.* **A42** (1990) 38.

[22] E.J. Squires, in: *Quantum Chaos–Quantum Measurement*, P. Cvitanovic *et al.*, eds. (Kluwer Academic Publishers, Dordrecht, 1992).

[23] J.S. Bell, in: *The Ghost in the Atom*, P.C.W. Davies and J.R. Brown, eds. (Cambridge University Press, Cambridge, 1986).

[24] H.P. Stapp, in: *Philosophical Consequences of Quantum Theory*, J.T. Cushing and E. McMullin, eds. (University of Notre Dame Press, Notre Dame, Indiana, 1989).

[25] N.D. Mermin, *Journ. Phil.* **78** (1981) 397.

[26] A. Einstein, in *Albert Einstein: Philosopher-Scientist*, P.A. Schilpp, ed. (Tudor Pub. Co., New York, 1949).

[27] A. Shimony, in: *The New Physics*, P. Davies, ed. (Cambridge University Press, 1989).

[28] J.S. Bell, *Beables for Quantum Field Theory*, 1984, Aug. 2, CERN-TH. 4035/84.

[29] P. Suppes and M. Zanotti, in: *Logic and Probability in Quantum Mechanics*, P. Suppes, ed. (Reidel, Dordrecht, 1976); B. van Fraassen, *Synthese* **52** (1982) 25; J.P. Jarrett, *Nous* **18** (1984) 569; A. Shimony, in: *Proc. Int. Symp. on the Foundations of Quantum Mechanics*, S. Kamefuchi *et al.*, eds. (Physical Society of Japan, Tokyo, 1983).

[30] G.C. Ghirardi and R. Grassi, *Stud. Hist. Phil. Sci.* **25** (1994) 397.

[31] P. Eberhard, *Nuovo Cimento* **46B** (1978) 392; G.C. Ghirardi, A. Rimini and T. Weber, *Lett. Nuovo Cimento* **27** (1980) 293; G.C. Ghirardi, R. Grassi, A. Rimini and T. Weber, *Europhy. Lett.* **6** (1988) 95.

[32] G.C. Ghirardi, R. Grassi, J. Butterfield and G.N. Fleming, *Found. Phys.* **23** (1993) 341; J. Butterfield, G.N. Fleming, G.C. Ghirardi and R. Grassi, *Int. J. Theor. Phys.* **32** (1993) 2287; G.C. Ghirardi and R. Grassi, in: *Bohmian Mechanics and Quantum Theory: An Appraisal*, J. Cushing *et al.*, eds. (Kluwer Academic Publishers, Dordrecht, 1996).

[33] P. Pearle, in: *Sixty-two Years of Uncertainty*, A. Miller, ed. (Plenum, New York, 1990).

[34] G.C. Ghirardi, R. Grassi and P. Pearle, in: *Symposium on the Foundations of Modern Physics 1990*, P. Lahti and P. Mittelstaedt, eds. (World Scientific, Singapore, 1990).

[35] P. Pearle, in: *Quantum Chaos–Quantum Measurement*, P. Cvitanovic *et al.*, eds. (Kluwer Academic Publishers, Dordrecht, 1992).

[36] G.C. Ghirardi, *Found. Phys. Lett.* **9** (1996) 313.

[37] D. Dürr, S. Goldstein and N. Zanghí, *Journ. Stat. Phys.* **67** (1992) 843; in: *Bohmian Mechanics and Quantum Theory: An Appraisal*, J. Cushing *et al.*, eds. (Kluwer Academic Publishers, Dordrecht, 1996).

[38] G.C. Ghirardi and R. Grassi, in: *Nuovi Problemi della Logica e della Filosofia della Scienza*, D. Costantini *et al.*, eds. (Editrice Clueb, Bologna, 1991).

[39] T. Weber, *Nuovo Cimento* **106B** (1991) 1111.

[40] F. Benatti, G.C. Ghirardi, A. Rimini and T. Weber, *Nuovo Cim.* **101B** (1988) 333.

[41] L. Diosi, *Phys. Rev.* **A40** (1989) 1165.

[42] G.C. Ghirardi, R. Grassi and A. Rimini, *Phys. Rev.* **A42** (1990) 1057.

[43] A. Shimony in: *PSA 1990*, **2**, A. Fine, M. Forbes and L. Wessels, eds. (Philosophy of Science Association, East Lansing, Michigan, 1990).

[44] D.Z. Albert and L. Vaidman, *Phys. Lett.* **A139** (1989) 1; D.Z. Albert and L. Vaidman, in: *Bell's Theorem, Quantum Theory and Conceptions of the Universe*, M. Kafatos, ed. (Kluwer Academic Publ., Dordrecht, 1989); D.Z. Albert, in: *Sixty-two Years of Uncertainty*, A. Miller, ed. (Plenum, New York, 1990); D.Z. Albert and B. Loewer, in: *PSA 1990*, **1**, A. Fine, M. Forbes and L. Wessels, eds. (Philosophy of Science Association, East Lansing, Michigan, 1990); D. Albert, *Quantum Mechanics and Experience* (Harvard University Press, Cambridge, Mass., 1992).

[45] F. Aicardi, A. Borsellino, G.C. Ghirardi and R. Grassi, *Found. Phys. Lett.* **4** (1991) 109.

[46] G.C. Ghirardi, R. Grassi and F. Benatti, *Found. Phys.* **25** (1995) 5; G.C. Ghirardi, *Erkenntnis* **45** (1997) 349; G.C. Ghirardi and T. Weber, in: *Potentiality, Entanglement and Passion-at-a-Distance*, R.S. Cohen *et al.*, eds. (Kluwer Academic Publishers, Dordrecht, 1997); G.C. Ghirardi, in: *Structures and Norms in Science*, M.L. Dalla Chiara *et al.*, eds. (Kluwer Academic Publishers, Dordrecht, 1997).

5 Quantum effects in accelerator physics*

JON MAGNE LEINAAS

Quantum effects for electrons in a storage ring are discussed, in particular the polarization effect due to spin-flip synchrotron radiation. The electrons are treated as a simple quantum mechanical two-level system coupled to the orbital motion and the radiation field. The excitations of the spin system then are related to the Unruh effect, i.e. the effect that an accelerated radiation detector is thermally excited by vacuum fluctuations. The chapter reviews an earlier work which was done in collaboration with John Bell.

1 Introduction

I would like to begin by reminding you about John Bell's strong connection to the field of accelerator physics. In fact he started out his physics career in this field, and although his main contributions have been within other parts of theoretical physics, he has through the years written a number of papers related to this subject. Many of these have been written together with his wife, Mary Bell. Let me just mention as key words: from his early period, works on linear accelerators and the physics of strong focusing machines; from later years, papers on electron cooling in storage rings [1] and on radiation damping [2], on spin and polarization effects of stored electrons [3, 4, 5] and on the phenomenon of quantum beamstrahlung [6]. On the last of these subjects he wrote, with Mary Bell, several papers in the last few years of his life.

My own contact with this field is only through the work I did with John some years back [4, 5]. This started when I was at CERN as a fellow almost ten years ago. I was then interested in something quite different from accelerator physics,

*From an invited talk at the Symposium on Quantum Physics at CERN, May 2nd and 3rd 1991, in memory of John S. Bell.

namely in the rather exotic theoretical effect often referred to as the Unruh effect. As shown by Unruh [7], an idealized radiation detector which is accelerated through ordinary Minkowski vacuum gets heated due to interactions with the vacuum fluctuations of the radiation field. For uniform linear acceleration the excitation spectrum has a universal, thermal form, independent of details of the detector. This effect, that vacuum seems hot, as measured in an accelerated system, has by Unruh and others been related to the phenomenon of Hawking radiation from black holes [8].

I discussed with John Bell, whom I had got to know at that time, whether it would be possible to see this effect in real experiments, or whether the strong accelerating forces needed would make it impossible, even in principle, to make a sufficiently robust detector. I was inclined to think so and had some arguments in that direction. But John then got the idea that a depolarization effect which was known to exist for electrons in a storage ring could have something to do with the Unruh heating. We investigated that and found that the effect indeed was related, although there were important complications due to the fact that the electrons were following a circular orbit rather than being linearly accelerated [4]. Motivated by these complications we later examined more carefully the effects of quantum fluctuations for an electron moving in a circular orbit [5].

In this chapter I will discuss the polarization effects for circulating electrons, mainly in the way presented in our two papers [4, 5]. However, I begin by focusing attention on the quantum description of electrons in a storage ring, and only later relate this to the Unruh effect. I will consider the quantum effects of the accelerated electrons under simple, idealized conditions, and avoid many complications which are relevant and important for electrons in real accelerators.

I would like to add in this introduction that for me it was a highly inspiring experience to work with John Bell. He had a way of reaching the essence of the problem under discussion, and of avoiding all unnecessary complications, which I found both remarkable and challenging. But more generally than this I was much attracted by John Bell's way of understanding and of doing physics. In addition he was a very pleasant person. I always liked very much to see him, as I have done from time to time also in recent years, both to discuss physics with him and also to hear his views on other matters.

2 Quantum effects for accelerated electrons

The motion of particles in accelerators can mostly be understood and described in classical terms. But there are some quantum effects which are non-negligible and which even may be important. These mainly have to do with radiation phenomena and with the radiation reaction on the accelerated particles. Therefore they are much more important for the light electron than for the much heavier proton. For this reason I shall restrict myself to discussing quantum effects for accelerated electrons.

The accelerated electrons emit radiation, synchrotron radiation, as it is known for particles in a magnetic field. Even for high energy electrons this process is well described by the classical radiation formula. This was explicitly demonstrated by Schwinger [9] who calculated the lowest order quantum correction to the radiated power. Only for extremely high energetic electrons do the quantum corrections become important. The condition for this being small can be written as

$$\gamma \ll \gamma_c = \sqrt{\frac{mc\rho}{\hbar}}, \qquad \gamma = \frac{1}{\sqrt{1-\left(\frac{v}{c}\right)^2}}, \tag{1}$$

with m as the electron mass, v its velocity and ρ the radius of curvature of the particle orbit. The important ratio then is

$$\Upsilon = \left(\frac{\gamma}{\gamma_c}\right)^2 \approx \frac{a}{a_m} \qquad (\gamma \gg 1), \tag{2}$$

where a is the acceleration of the particle in an inertial rest frame, and a_m is an acceleration parameter determined by the particle mass,

$$a = \frac{\gamma^2 v^2}{\rho}, \qquad a_m = \frac{mc^3}{\hbar}. \tag{3}$$

Thus the important physical quantity is the acceleration a rather than the energy of the electron. A typical value for the parameter Υ in cyclic accelerators is 10^{-6}, which shows that quantum effects indeed are very small.

However, recently there has been some interest in the quantum regime $\Upsilon > 1$ in connection with linear accelerators at very high energy [10, 11, 12]. In linear colliders with energies much above those of machines of today such high accelerations may be produced when electrons pass through a bunch of positrons, and vice versa. A characteristic feature of the radiation process then is that the emitted photon takes a large fraction of the electron's energy. This phenomenon has been referred to as quantum beamstrahlung. As already mentioned, some of the last papers of John Bell were dealing with the description of this effect [6]. However, I will not discuss this effect, but rather consider quantum effects for much smaller accelerations, $a \ll a_m$.

Even if synchrotron radiation is essentially a classical phenomenon for $\gamma \ll \gamma_c$, this does not mean that all quantum effects associated with this phenomenon are unimportant. The radiation reaction will excite orbital oscillations even for much smaller energies [13]. The radiation field then acts in two ways on the particle. Quantum fluctuations excite the oscillations, while radiation damping tends to reduce the oscillation amplitudes. Balance between these two tendencies defines a minimal, quantum limit to the beam size.

However, a perhaps even clearer demonstration of quantum effects for electrons in a storage ring is associated with the phenomenon of spin–flip radiation [14, 15]. The asymmetry between up and down flips in the magnetic field leads to a gradual build up of transverse polarization of the electrons. Under ideal conditions the polarization approaches equilibrium as

$$P(t) = P_0(1 - \mathrm{e}^{-t/t_0}), \tag{4}$$

with a maximum polarization

$$P_0 = \frac{8}{5\sqrt{3}} = 0.924 \tag{5}$$

and a build up time

$$t_0 = \frac{8}{5\sqrt{3}} \frac{m^2 c^2 \rho^3}{\mathrm{e}^2 \hbar \gamma^5}\,. \tag{6}$$

For existing accelerators this build up time is of the order of minutes to hours.

The phenomenon of spontaneous polarization of electrons circulating in a magnetic field has been analyzed in many publications, both for ideal conditions and for the more realistic situation with particles moving in a variable magnetic field. There also exist review articles on this interesting subject [16, 17, 18]. In particular the paper by J.D. Jackson focuses attention on the aspects of this phenomenon which can be described in elementary terms. My approach will be along the same lines. But whereas Jackson rejects the idea of a simple description of the effect as a transition between spin energy levels caused by radiation effects, which then would lead eventually to all particles in the lowest energy level, this is exactly the picture I will use. The electrons can be treated as a simple one-particle quantum system interacting with the radiation field. But the effect of the radiation field along the accelerated orbit of the electrons is different from the effect on an electron sitting at rest. Transitions also to the upper energy level are induced by the field along this orbit, and that leads to a small but non-vanishing depolarization of the electron beam. I will only consider the ideal case of electrons moving in a rotationally invariant magnetic field. A stable, classical circular orbit is produced by a radial gradient in the magnetic field. This corresponds to the situation of a weak focusing machine.

3 Quantum mechanics in the accelerated frame

Let me first remind you about some facts concerning the relativistic, classical description of spin motion. When the spin is described as a 3-vector we actually refer to the spin in the rest frame of the particle. But this vector can be included in the lab frame description of the particle by a boost between the two frames. This is the standard approach. Here I would like instead to stay in the rest frame, and to consider also the particle motion in this frame.

Actually there are three different co-moving frames which are characterized by simple properties, in one way or the other. The first one, which I will denote the L-frame, is the one which is obtained from the fixed lab frame by a pure boost. This frame is non-rotational as seen from the lab. The other one is the frame which rotates with the frequency of the orbiting particle. In this frame, which I will denote the O-frame, the accelerating field is stationary. Finally there is a frame, denoted the C-frame, which is non-rotational along the particle orbit. The fact that this is different from the L-frame is a relativistic effect and gives rise to Thomas precession of the spin vector. The relative rotational frequencies of these frames are listed in Table 5.1 for an electron following a circular path in a magnetic field. Both the magnetic field B and the frequencies refer to co-moving frames.

Also the spin precession frequencies in the three different frames are shown in Table 5.1. This has the simplest form in the C-frame. Since this frame is non-rotational along the orbit, all spin precession in this frame is due to coupling

Table 5.1. *Rotation frequencies of three different co-moving frames and the spin precession frequencies in these frames for an electron orbiting in a constant magnetic field B*

Frames	Rotational freq.	Spin precession
C	0	$g\dfrac{e}{2mc}B$
O	$\dfrac{e}{mc}B$	$(g-2)\dfrac{e}{2mc}B$
L	$\left(1-\dfrac{1}{\gamma}\right)\dfrac{e}{2mc}B$	$\left((g-2)+\dfrac{2}{\gamma}\right)\dfrac{e}{2mc}B$

between the magnetic moment of the particle and the external magnetic field. Thus, the precession frequency is proportional to the gyromagnetic factor g, as shown in the table. In the O-frame, on the other hand, the precession frequency is proportional to $g-2$. This demonstrates the well-known fact that for $g=2$ the spin precesses exactly with the frequency of the orbital motion. Finally, in the L-frame there is a further correction due to the relative rotation of the L- and the O-frame. This correction is identical to the rotation frequency of the orbital motion, and this is smaller than the frequency associated with the Thomas precession by a factor of $1/\gamma$.

Even if the spin motion is simplest in the C-frame, I shall in the following apply the O-frame for the quantum description. The reason for this is that the external fields are stationary in this frame and therefore give rise to a time-independent Hamiltonian. This frame will be extended to a local accelerated coordinate system to allow for fluctuations in the particle about the circular path, which then is assumed to be the classical, stable orbit of the electrons. This coordinate system will necessarily contain coordinate singularities at some distance from the orbit. But I will assume fluctuations away from the stable orbit to be small, so that linearized equations are sufficient. I will also assume velocities in this frame to be small, so that a non-relativistic approximation can be applied.

The Hamiltonian which governs the time evolution in the accelerated frame is not identical to that of the inertial rest frame, but it can be expressed in a simple way in terms of observables from this frame,

$$H' = H - \frac{a}{v}J_z + \frac{a}{c}K_x. \tag{7}$$

Here H is the Dirac Hamiltonian, J_z is the generator of rotations in the plane of the electron orbit and K_x is a boost operator. The coordinates in the O-frame then are chosen with the particle acceleration in the negative x-direction and with the orbit velocity in the positive y-direction. The additional terms in the expression for H' are fairly easy to understand. The presence of the generator of rotations is related to the fact that the coordinate axes of the O-frame rotate along the orbit and the presence of the boost operator accounts for the continuous jumping between inertial frames when the particle is accelerated. For the three operators included in H' we have the following expressions

$$H = c\vec{\alpha}\cdot\vec{\pi} + \beta mc^2 + \mathrm{e}\phi + \kappa\frac{\mathrm{e}\hbar}{2mc}(\mathrm{i}\beta\vec{\alpha}\cdot\vec{E} - \beta\vec{\sigma}\cdot\vec{B}), \tag{8}$$

$$J_x = (\vec{r}\times\vec{p})_z + \frac{1}{2}\hbar\sigma_z, \tag{9}$$

$$K_x = -\frac{1}{2c}(xH + Hx), \tag{10}$$

where a term for the anomalous magnetic moment, $\kappa = \frac{1}{2}(g-2)$, has been introduced in the expression for H. All the potentials and fields in these expressions refer to the inertial rest frame of the classical orbit. The notation $\vec{\pi} = \vec{p} - (\mathrm{e}/c)\vec{A}$ has been used for the mechanical momentum of the electron.

When the fluctuations around the classically stable orbit are assumed to be small, then a non-relativistic approximation makes sense. A Foldy–Wouthuysen transformation, where we keep only the leading terms, gives a Hamiltonian which then can be split into a spin-independent and a spin-dependent term in the following way:

$$H = H_{\mathrm{orb}} + H_{\mathrm{spin}} \tag{11}$$

$$H_{\mathrm{orb}} = \left(1 - \frac{ax}{c^2}\right)\left(\frac{\vec{\pi}^2}{2m} + \mathrm{e}\phi\right) - axm + \frac{\mathrm{i}a\hbar}{2mc^2}\pi_x - \frac{a}{v}(\vec{r}\times\vec{p})_z + \cdots \tag{12}$$

$$\begin{aligned} H_{\mathrm{spin}} = &-\frac{\mathrm{e}\hbar}{4mc}\vec{\sigma}\cdot\left[\left(1 - \frac{ax}{c^2}\right)g\vec{B} + \frac{g-1}{mc}\vec{E}\times\vec{\pi}\right] \\ &- \frac{1}{2}\frac{a\hbar}{v}\sigma_z + \frac{a\hbar}{4mc^2}(\vec{\sigma}\times\vec{\pi})_x + \cdots \end{aligned} \tag{13}$$

The Hamiltonian in the accelerated frame includes some complications compared to that of the inertial frame. However, if we consider the orbital motion only to linear order in the deviation from the stable orbit, the main difference in the expressions for H_{orb} is the presence of the centrifugal and Coriolis terms. For the orbital motion the spin effects only give rise to a small perturbation. But also for the spin motion the effect of the fluctuations in the orbit is small, since the spin precession is mainly determined by the strong magnetic field along the classical orbit. In principle one could then determine the particle motion in the following way: one first solves for the orbital motion, neglecting the spin, and then one determines the spin motion, treating the orbital fluctuations and the quantum fields as perturbations. However, when calculating the polarization of the particle beam, one can simplify this approach somewhat, since it is only the fluctuations in the particle orbit which are driven by the coupling to the radiation field which are important.

4 Spin transitions

The spin Hamiltonian can be written in the form

$$H_{\text{spin}} = \frac{1}{2}\hbar\vec{\omega}\cdot\vec{\sigma} \tag{14}$$

$$\vec{\omega} = \vec{\omega}_0 + \delta\vec{\omega}, \tag{15}$$

with $\vec{\omega}_0$ giving rise to the classical part of the precession,

$$\vec{\omega}_0 = -\frac{\mathrm{e}}{2mc}g\vec{B}_0 - \frac{a}{v}\vec{k} = -\frac{\mathrm{e}}{2mc}(g-2)B_0\vec{k}, \tag{16}$$

and $\delta\vec{\omega}$ as the fluctuation part,

$$\delta\vec{\omega} = -\frac{\mathrm{e}}{2mc}\left[g\delta\vec{B} - \frac{ax}{c^2}g\vec{B}_0 - (g-2)\frac{a}{\mathrm{e}c}\vec{i}\times\vec{\pi}\right]. \tag{17}$$

In the last two expressions $\vec{k}$ is the unit vector orthogonal to the plane of motion and $\vec{B}_0 = B_0\vec{k}$ is the external magnetic field along the classical orbit. $\delta\vec{B}$ accounts for the fluctuations in the magnetic field. This can be separated into two parts,

$$\delta\vec{B} = \vec{B}_{\text{q}} + \delta\vec{B}_{\text{c}}, \tag{18}$$

where $\vec{B}_{\text{q}}$ denotes the quantum field along the classical orbit and $\delta\vec{B}_{\text{c}}$ is the variation in the external field due to fluctuations in the orbit.

The spin motion now can be determined by time dependent perturbation theory. $\vec{\omega}_0$ then defines the unperturbed part of the spin Hamiltonian and $\delta\vec{\omega}$

the perturbation. To first order, the transition probabilities per unit time between the levels of the unperturbed Hamiltonian are given by

$$\Gamma_\pm = \lim_{T\to\infty} \frac{1}{4T} \left| \int_{-T/2}^{T/2} \mathrm{e}^{\pm \mathrm{i}\omega_0 \tau} \delta\omega_\mp(\tau)|0\rangle \right|^2$$
$$= \frac{1}{4} \int_{-\infty}^{+\infty} \mathrm{d}\tau\, \mathrm{e}^{\mp \mathrm{i}\omega_0\tau} \langle 0|\delta\omega_\pm(\tau/2)\delta\omega_\mp(-\tau/2)|0\rangle. \tag{19}$$

$|0\rangle$ in this equation denotes the state of the combined system of radiation field and orbit variables, unperturbed by the spin. $\delta\omega_\pm$ is a linear combination of the x- and y-component of $\delta\vec{\omega}$,

$$\delta\omega_\pm = \delta\omega_x \pm \mathrm{i}\,\delta\omega_y. \tag{20}$$

The same notation will be used for other variables in the following.

A useful substitution rule which can be used in the expression for $\delta\omega_\pm$ is the following one:

$$\frac{\mathrm{d}}{\mathrm{d}\tau} F \to \pm \mathrm{i}\omega_0 F. \tag{21}$$

The difference between these two expressions only gives rise to end effects in the integral for the transition amplitude, and for large T this is suppressed in $\Gamma_\pm$ due to the prefactor $1/T$. This substitution rule can now be used to eliminate the orbital variables in the expression for $\delta\omega_\pm$, which can be written in the form

$$\delta\omega_\pm = -\frac{\mathrm{e}}{2mc}\left[gB_{\mathrm{q}\pm} + g\delta B_{\mathrm{c}\pm} \pm 2\mathrm{i}v\omega_0 \frac{m}{\mathrm{e}c}\dot{z} \right]. \tag{22}$$

With the stable orbit in the symmetry plane of the magnetic field, $\delta B_{\mathrm{c}\pm}$ gets contribution only from the gradient in the z-direction. This implies that (to lowest order) $\delta\omega_\pm$ depends only on the vertical fluctuations in the particle orbit. These fluctuations in turn are determined by coupling to the radiation field in the following way

$$\ddot{z} - \frac{2\mathrm{e}^2}{3mc^2}\left(\dddot{z} - \frac{a^2}{c^2}\dot{z} \right) + \Omega^2 z = \frac{\mathrm{e}}{m} E_{\mathrm{q}z}, \tag{23}$$

where a radiation reaction term has been introduced and where nonlinear terms have been neglected. The restoring electric force in the z direction can be related to the gradient in the magnetic field,

$$\Omega^2 = \frac{a}{\rho} n, \tag{24}$$

$$n = \frac{\rho}{B_0}\frac{\partial B_z}{\partial x} = \frac{\rho}{B_0}\frac{\partial B_x}{\partial z}. \tag{25}$$

ρ is the radius of the (classical) electron orbit, and n the fall-off parameter of the magnetic field.

By use of the substitution rule, Eq. (21) can now be solved for z,

$$z = [\Omega^2 - \omega_0^2 \pm \mathrm{i}\Delta\omega_0]^{-1} \frac{\mathrm{e}}{m} E_{\mathrm{q}z} + \Lambda, \tag{26}$$

$$\Delta = \frac{2\mathrm{e}^2}{3mc^3}\left(\frac{a^2}{c^2} + \omega_0^2\right). \tag{27}$$

Λ here denotes a term which is suppressed for large T. When the expression for z is inserted in Eq. (22), this gives (for $v \approx c$),

$$\delta\omega_\pm = -\frac{\mathrm{e}}{2mc}[gB_{\mathrm{q}\pm} + (2 + f_\pm(g))E_{\mathrm{q}z}], \tag{28}$$

with $f_\pm(g)$ as a resonance term,

$$f_\pm(g) = \frac{(g-2)\Omega^2}{\Omega^2 - \omega_0^2 \pm \mathrm{i}\Delta\omega - 0}. \tag{29}$$

This term blows up when the frequency of the free oscillations in the z-direction is close to the classical spin precession frequency, but it tends rapidly to zero away from the resonance.

The new expression for $\delta\omega_\pm$ (28) now only depends on the free quantum fields, and the transition probabilities can be expressed in terms of correlation functions of these fields along the particle orbit,

$$\Gamma_\pm = \frac{1}{4}\int_{-\infty}^{+\infty} \mathrm{d}\tau\, \mathrm{e}^{\mp\mathrm{i}\omega_0\tau} \langle 0|\delta\omega_\mp(\tau/2)\delta\omega_\pm(-\tau/2)|0\rangle. \tag{30}$$

(A correct treatment of the singularity at $\tau = 0$ corresponds to a small shift $\tau \to \tau - \mathrm{i}\varepsilon$.) $|0\rangle$ then refers to the vacuum state of the radiation field. To calculate these probabilities now is straightforward. The fields in the co-moving frame are most conveniently expressed in terms of lab frame fields, and the correlation functions of these are found by expressing the field operators in terms of creation and annihilation operators. To leading order in $1/\gamma$ the

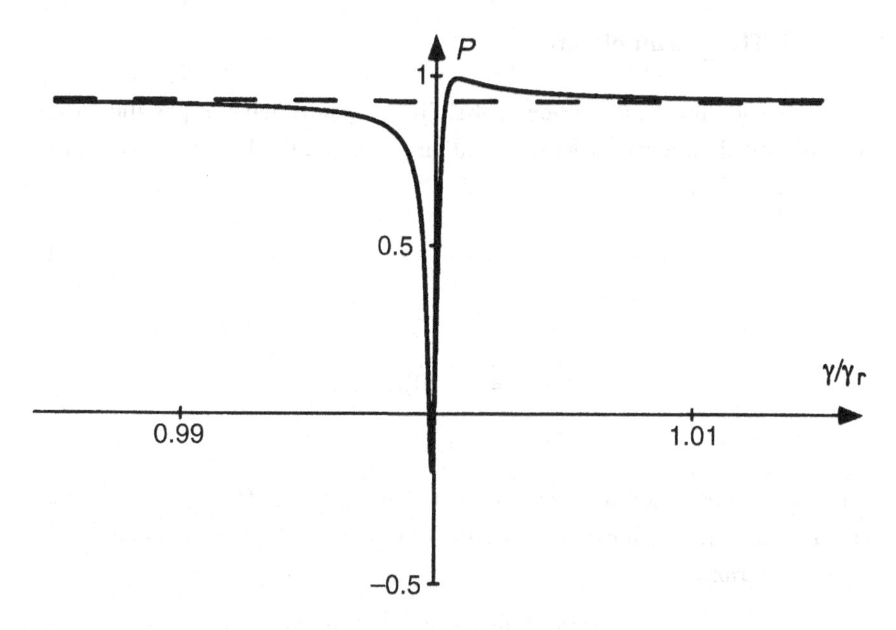

FIGURE 5.1 The equilibrium polarization P as a function of γ close to the resonance with the vertical oscillations. The dashed line corresponds to the limiting value $P = 0.92$, away from the resonance. The scale for γ is relative to the resonance value γ_r. Positive polarization corresponds to the direction opposite to the magnetic field.

relevant Fourier integrals can be calculated analytically. I will not discuss details about this here, but only show the result for the stationary value of the polarization, Fig. 5.1. The polarization is determined by the population of the two spin levels, and this in turn is found by the standard argument of equilibrium between transitions up and down. We have

$$P = \frac{\Gamma_+ - \Gamma_-}{\Gamma_+ + \Gamma_-}. \tag{31}$$

In Fig. 5.1 the polarization is shown as a function of γ. Except for values close to the resonance with the vertical motion, the standard result for the polarization is found, $P = 0.924$. The effect of the resonance is mainly to depolarize the beam, but an interesting detail is the coherent effect which gives a maximum value of $P = 0.992$ close to the resonance. Thus, at least in principle, it is possible to exceed the limiting value of 0.924.

5 The Unruh effect

The expression for the transition probabilities $\Gamma_\pm$ now makes it possible to see the close relation between the polarization effect and the Unruh effect [7]. Let me rewrite it in the form

$$\Gamma_\pm = \int_{-\infty}^{+\infty} \mathrm{e}^{\mathrm{i}\omega_0\tau} C_\pm(\tau), \tag{32}$$

with

$$C_+(\tau) = \langle D^\dagger(0) D(\tau)\rangle, \tag{33}$$

$$C_-(\tau) = \langle D(\tau) D^\dagger(0)\rangle. \tag{34}$$

I have here introduced the new notation $D = (1/2)\delta\omega_+$, $D^\dagger = (1/2)\delta\omega_-$. The operator D is then a linear combination of electric and magnetic fields in the co-moving frame,

$$D(\tau) = \vec{\alpha}\cdot\vec{E}(x(\tau)) + \vec{\beta}\cdot\vec{B}(x(\tau)), \tag{35}$$

where $x(\tau)$ is the particle orbit. The expression (32) is similar to that which defines transitions in a point detector in the case of the Unruh effect. The main difference is that the world line of the detector then corresponds to linear acceleration rather than to circular motion as in the present case. However, the excitations of the accelerated systems in the two cases can be understood qualitatively in the same way. The correlation functions $C_\pm$ give a measure of vacuum fluctuations of the electromagnetic field along the orbit $x(\tau)$, and these fluctuations give rise to excitations in the detector when C_+ includes a spectral component which coincides with the excitation energy.

To see this more clearly, let me first discuss the simplest case, namely with a two level detector at rest. The transitions between the levels are given by the same set of equations, (32–35), but now simply with

$$x(\tau) = (\tau, 0). \tag{36}$$

Since $D(\tau)$ is a linear combination of electromagnetic fields it can be decomposed in the form

$$D(\tau) = \int \mathrm{d}^3k \sum_r \left[c_r(\vec{k})\, \mathrm{e}^{-\mathrm{i}kx} a_r(\vec{k}) + d_r(\vec{k})\, \mathrm{e}^{+\mathrm{i}kx} a_r^\dagger(\vec{k})\right], \tag{37}$$

where $a_r(\vec{k})$ and $a_r^\dagger(\vec{k})$ are photon annihilation and creation operators and $c_r(\vec{k})$ and $d_r(\vec{k})$ are fourier coefficients. According to Eq. (32) it is only the positive frequency parts of this operator which are relevant for the transitions. Positive frequency is then measured relative to the proper time along the orbit $x(\tau)$. But with the detector at rest this coincides with positive frequencies measured in the lab frame. And, as is well known, the positive frequency part of lab frame fields contains only annihilation operators. This is simply because all excitations in the lab frame have positive energy. So the relevant component of $D(\tau)$ is

$$\int_{-\infty}^{+\infty} \mathrm{d}\tau \mathrm{e}^{\mathrm{i}\omega_0\tau} D(\tau) = 2\pi \int \mathrm{d}\Omega_k \omega_0^2 \sum_r c_r(\vec{k}) \mathrm{e}^{\mathrm{i}\vec{k}\cdot\vec{x}} a_r(\vec{k}), \tag{38}$$

which only annihilates photons with the same energy as the energy splitting of the two-level system. As a consequence of this the probability for transitions up in energy is zero, since the D operator then acts on the vacuum state. However, transitions down may be different from zero, since in this case it is instead $D^\dagger$ which acts on the vacuum state. Thus, the vacuum fluctuations only induce transitions to lower energies. This clearly is related to energy conservation in the combined system of detector and radiation field.

If the two-level system moves with constant velocity the picture is the same, since the sign of the zero component of the photon momentum k is the same in all inertial frames. The only way to have a non-zero probability for excitations to higher energies is to include other states than the vacuum state in the one which the operator D acts on. In particular the probability is non-zero for states with temperature $T \neq 0$.

However, for accelerated motion this is no longer the case. $x(\tau)$ then is no longer a linear function of τ and both functions $\mathrm{e}^{-\mathrm{i}kx(\tau)}$ and $\mathrm{e}^{+\mathrm{i}kx(\tau)}$ will in general have positive frequency parts in terms of the variable τ. In addition, for the electromagnetic field, there will be a τ-dependent Lorentz transformation connecting the fields in the co-moving frame with the lab frame fields. The net effect is to introduce a mixing between the positive and negative frequency parts, so that both the annihilation and the creation parts of the operator D will have positive frequency components in terms of the time variable τ. As a consequence of this there will be in general non-vanishing probabilities for excitations both up and down in energy for the accelerated system, even with the quantum field in the vacuum state.

For uniform linear acceleration along the z-axis, the accelerated path $x(\tau)$ is described by

$$t = \frac{a}{c}\sinh\left(\frac{a}{c}\tau\right), \qquad z = \frac{a}{c}\cosh\left(\frac{a}{c}\tau\right), \qquad x = y = 0. \tag{39}$$

The trajectory $x(\tau)$ in this case depends only on one free parameter, which is the rest frame acceleration a. An interesting symmetry which is present for this motion corresponds to a shift in the τ-parameter in the imaginary direction,

$$x(\tau) = x\left(\tau + \mathrm{i}\frac{2\pi c}{a}\right). \tag{40}$$

This symmetry, together with general symmetries from field theory, related to PCT invariance ([19, 20, 21]), gives a simple relation between the correlation functions corresponding to transitions up and down in energy,

$$C_+(\tau) = C_-\left(\tau - \mathrm{i}\frac{2\pi c}{a}\right). \tag{41}$$

This relation is similar to one which is present for correlation functions at non-zero temperature, and it leads to a similar result for the ratio between probabilities for transitions up and down,

$$\Gamma_+ = \int_{-\infty}^{+\infty} \mathrm{d}\tau\, \mathrm{e}^{\mathrm{i}\omega_0\tau} C_-\left(\tau - \mathrm{i}\frac{2\pi c}{a}\right) \tag{42}$$

$$= \int_{-\infty}^{+\infty} \mathrm{d}\tau\, \mathrm{e}^{\mathrm{i}\omega_0(\tau + \mathrm{i}2\pi c/a)} C_-\left(\tau - \mathrm{i}\frac{2\pi c}{a}\right) \tag{43}$$

$$= \exp\left(-\frac{\hbar\omega_0}{a\hbar/2\pi c}\right)\Gamma_-. \tag{44}$$

If the ratio between the two transition probabilities is now interpreted as a Boltzmann factor, then there is a simple linear relation between the temperature associated with this factor and the acceleration a,

$$kT_{\mathrm{U}} = \frac{a\hbar}{2\pi c}. \tag{45}$$

T_{U} is then the Unruh temperature for the accelerated system and k the Boltzmann constant. The derivation shows that the thermal property of the excitation spectrum depends only on general properties of the quantum fields and on special properties of the accelerated trajectory $x(\tau)$. Details of the accelerated system are not important. Returning to the case of accelerated electrons, one then might be tempted to consider the case of linear acceleration as the best case for seeing the Unruh effect in a real experiment. An additional magnetic field along the electron path could provide the necessary splitting of the spin energy levels. However, as discussed in ref. [4], the time needed to reach equilibrium is far too long to make this relevant for the motion of electrons in linear accelerators. For electrons in cyclic accelerators much larger accelerations can be obtained and correspondingly much smaller time constants for the approach to equilibrium.

Let me now briefly return to the case of the circulating electrons to look at the electron polarization from the point of view of the Unruh effect. One main difference between circular motion and linear acceleration is that the former depends on two independent parameters, which we may take to be a and γ. If we disregard this complication and naively assume that the acceleration a is the important one, in the same way as for the linear Unruh effect, then we find the following. The energy splitting of the spin system is for $\gamma \gg 1$,

$$\Delta E = \hbar\omega_0 = (g-2)\frac{\hbar a}{2c}. \tag{46}$$

This gives a Boltzmann factor

$$\exp\left(-\frac{\Delta E}{kT_{\mathrm{U}}}\right) = \exp[-\pi(g-2)], \tag{47}$$

where T_{U} is the Unruh temperature given by (45). Numerically, the factor is close to one, 0.99, as compared to the correct value for the ratio between the population of the two spin levels (away from the resonance), which is close to zero, 0.04. So a naive application of the Unruh temperature formula to the case of electrons accelerated in circular orbit does not give the correct value for the population of the two spin levels, and therefore for the polarization.

However, it may still be of interest to make a closer comparison of the two effects. Instead of considering only the physical value of g we may consider the functional dependence of g for the relative population of the two spin levels. We first note that the correct ratio $R(g)$ in fact can be written as a function of $\Delta E/kT_{\mathrm{U}}$, but it is not a simple exponential function. Nevertheless, a plot of $R(g)$ and the Boltzmann factor (47) shows a clear resemblance between these two functions (Fig. 5.2). The main difference is a relative shift along the g-axis. In

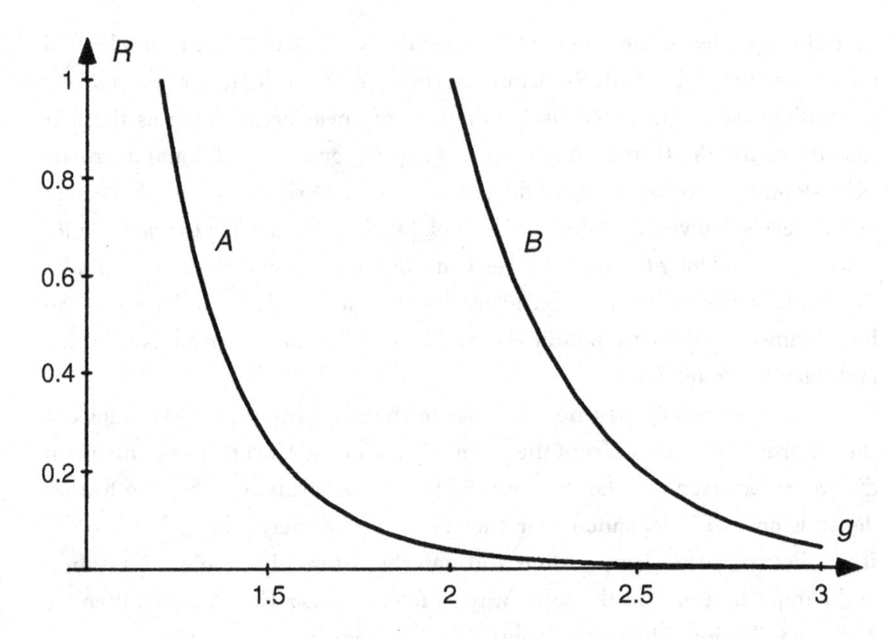

FIGURE 5.2 The ratio $R(g)$ between the equilibrium population of the two spin levels as a function of the gyromagnetic factor g. A is the result of the detailed calculation. B follows from assuming a thermal distribution over the levels in the O-frame, with temperature dertermined by the Unruh formula.

fact, such a complication is not totally unexpected, since the Boltzmann factor (47) refers to the rotating O-frame. If the excitation spectrum were to have a thermal form it might be more natural to assume this to be the case in the non-rotational C-frame. That would give a shifted exponential curve, but shifted with two units along the g-axis. The correct curve is located somewhere between these two exponential curves, and I have no simple explanation for the exact position of this curve. For large g-factors the rotation of frames is less important, and the ratio $R(g)$ then in fact is exponentially damped,

$$R(g) \approx \frac{0.016}{g} \exp[-\sqrt{3}(g-2)]. \tag{48}$$

This corresponds to a temperature which is somewhat higher than the Unruh temperature,

$$T_{\text{eff}} \approx \frac{\pi}{\sqrt{3}} T_{\text{U}} \approx 1.8\, T_{\text{U}}. \tag{49}$$

I will now leave the question of polarization of the accelerated electrons and as a final point briefly return to the orbital excitations. The (Heisenberg) equation of motion for the vertical oscillations can be written as

$$\ddot{z} + 2\Gamma\dot{z} + \Omega^2 z = \frac{e}{m} E_{qz}. \tag{50}$$

In this equation $\Gamma = (e^2a^2)/(3mc^5)$ and only the most important part of the radiation damping term has been kept. Making use of the fact that the damping is small, $\Gamma \ll \Omega$, one can solve the equation to find an (approximate) expression for $z(\tau)$ in terms of the quantum field E_{qz}. For the fluctuation in the z-coordinate one finds the following expression:

$$\langle z^2 \rangle = \frac{1}{2\Gamma}\left(\frac{\mathrm{e}}{m\Omega}\right)^2 \int_{-\infty}^{+\infty} \mathrm{d}\tau\, \mathrm{e}^{-\Gamma|\tau|} \cos\Omega\tau \langle E_{qz}(\tau/2) E_{qz}(-\tau/2)\rangle. \tag{51}$$

This shows that the fluctuations in the vertical direction are determined by the correlation function of the z-component of the electric field along the classical orbit. The vertical fluctuations in fact can be interpreted as being due to the 'circular Unruh effect' in a similar way to the polarization effect. The mean energy associated with the fluctuations is, for large γ,

$$\langle E \rangle_{\text{vert}} = m\Omega^2 \langle z^2 \rangle = \frac{13}{96}\sqrt{3}\,\frac{a\hbar}{c}. \tag{52}$$

It is proportional to the acceleration a, but with a different prefactor as compared with the linear Unruh effect. it corresponds to a somewhat higher temperature

$$T_{\text{eff}} \approx 1.5\, T_{\text{U}}. \tag{53}$$

To linear order the excitation spectrum in this case in fact has a thermal form, and there is no complication with rotating frames. So in this respect the vertical orbit excitations give a simpler demonstration of the Unruh heating in the circular case than the depolarization of the electrons do. But the fluctuations are small and to measure them may be a much more difficult task. As a final comment let me just mention that the horizontal fluctuations are different. They are larger than the vertical fluctuations, essentially by a factor γ^2. These fluctuations then depend not only on the acceleration a of the electrons (for $\gamma \gg 1$), but on both the parameters a and γ which characterize the circular orbit.

6 Concluding remarks

In this chapter I have discussed quantum effects within a simple idealized model of a cyclic accelerator. In a more realistic case there will be several modifications of this picture. In the case of a strong focusing machine the magnetic field will no longer be uniform along the orbit. The unperturbed part of the spin Hamiltonian will then be time dependent, and as a consequence of this the perturbations cannot be described in terms of transitions between *stationary* spin levels. But the effect of the perturbations instead can be described as giving rise to transitions between *periodic orbits* in the spin variable.

In addition to this there may be other perturbations in the magnetic field that cause a coupling between vertical and horizontal oscillations. Also non-linear effects may be important. This will in general lead to a much richer structure of spin–orbit resonances than in the idealized model where only one resonance is present. All these effects certainly have to be taken into account when one wants to model the spin behaviour in a real accelerator. (For some recent references where complications of this kind are included, see [22, 23, 24].) Nevertheless, to understand the main aspects of the quantum effects for the accelerated electrons, such a simple, idealized model may be of interest. Let me therefore end by pointing to some of the features of this model which I want to stress.

The effect of spontaneous polarization of the circulating electrons, and the departure from full polarization, can be understood and described within a simple two-level model for the electrons. The external magnetic field along the orbit defines the (unperturbed) spin levels of the electrons, and the radiation field causes transitions between these two levels. The radiation field acts both directly on the spin, through the coupling to the magnetic moment, and also indirectly, through the fluctuations it introduces in the particle orbit.

Transition probabilities are determined by vacuum correlation functions of the electromagnetic fields along the classical orbit. No explicit reference to the radiation process is needed in this description. The effect then is similar to the Unruh effect for a linearly accelerated two-level system coupled to the radiation field. But there are complications due to the rotations of frames along the orbit.

The fluctuations in the orbital motion can be determined in a similar way. Vacuum fluctuations in the electromagnetic field along the classical orbit introduce orbital excitations. The vertical fluctuations have a thermal excitation spectrum, but with a slightly higher temperature than the Unruh temperature for linear acceleration.

Finally, in the simple model considered here, there is one resonance between spin and vertical oscillations. The main effect of the resonance is to depolarize the electron beam, but a detail which has been noted is the coherent effect which causes the polarization to pass the 'maximum' value of 0.92 close to the resonance.

REFERENCES

[1] J.S. Bell and M. Bell, *Particle Accelerators* **11** (1981) 233 and **12** (1982) 49.

[2] M. Bell and J.S. Bell, *Particle Accelerators* **13** (1983) 13.

[3] J.S. Bell, *CERN report* 75-11 (1975).

[4] J.S. Bell and J.M. Leinaas, *Nucl. Phys.* **B212** (1983) 131.

[5] J.S. Bell and J.M. Leinaas, *Nucl. Phys.* **B284** (1987) 488.

[6] M. Bell and J.S. Bell, *Particle Accelerators* **22** (1988) 301 and **24** (1988) 1, *Nucl. Instr. and Meth.* **A275** (1989) 258.

[7] W.G. Unruh, *Phys. Rev.* **D14** (1976) 870.

[8] S. Hawking, *Nature* **248** (1974) 30, *Comm. Math. Phys.* **43** (1975) 199.

[9] J. Schwinger, *Proc. Nat. Acad. Sci.* **40** (1954) 132.

[10] T. Himel and J. Siegrist, *AIP Conf. Proc. (USA)* no. 130, p. 602 (1985).

[11] R.J. Noble, *Nucl. Instr. and Meth.* **A256** (1987) 427.

[12] M. Jacob and T.T. Wu, *Phys. Lett.* **B197** (1987) 253.

[13] A.A. Sokolov and I.M. Ternov, *Synchrotron Radiation* (Akademie-Verlag, Berlin, 1968).

[14] A.A. Sokolov and I.M. Ternov, *Dokl. Akad. Nauk. SSR* **153** (1963) 1052 [*Sov. Phys. Dokl.* **8** (1964) 1203].

[15] Ya.S. Derbenev and A.M. Kondratenko, *Zh. Eksp. Teor. Fiz.* **64** (1973) 1918 [*Sov. Phys.-JETP* **37** (1973) 968].

[16] V.N. Baier, *Usp. Fiz. Nauk.* **105** (1971) 441 [*Sov. Phys.-Usp.* **14** (1972) 695].

[17] J.D. Jackson, *Rev. Mod. Phys.* **48** (1976) 417.

[18] B.W. Montague, *Physics Reports* **133** (1984) 1.

[19] G. Sewell, *Ann. Phys. (N.Y.)* **141** (1982) 201.

[20] R.J. Hughes, *Ann. Phys. (N.Y.)* **162** (1985) 1.

[21] J.S. Bell, R.J. Hughes and J.M. Leinaas, *Z. Phys.* **C28** (1985) 75.

[22] S.R. Mane, *Phys. Rev.* **A36** (1987) 105 and 120.

[23] J. Kewisch, R. Rossmanith and T. Limberg, *Phys. Rev. Lett.* **62** (1989) 419.

[24] L.N. Hand and A. Skuja, *Phys. Lett.* **A139** (1989) 291.

6 New aspects of Bell's theorem

ABNER SHIMONY

Bell's pioneering paper of 1964 [1] proved that all deterministic local hidden variables theories imply an inequality which is incompatible with some of the predictions of quantum mechanics. That paper raised many interesting questions, some of which have been measured thoroughly, some not completely, among them the following: is an assumption of determinism necessary for a derivation of Bell's Theorem? Can the locality assumption of Bell be analyzed and understood deeply? What quantum states imply a violation of Bell's Inequality? How decisively is Bell's Inequality refuted experimentally? Can Bell's Theorem be proved without an inequality? A review is given of some of the work on these and related questions, especially of some investigations of the last decade.[1]

1. Some history and non-history

The main historical event that I wish to recall is the publication by John Bell of a paper entitled 'On the Einstein–Podolsky–Rosen Paradox' [1], in volume 1 of the short-lived journal *Physics.* It contains the first formulation and proof of the theorem that no local hidden variables theory can be consistent with all the predictions of quantum mechanics – a theorem which has come to be known as 'Bell's Theorem'. This paper is very compact and establishes efficiently and elegantly what Bell wished to show. But it is also suggestive of further developments, some of which were made by him and some by others. This chapter will be an account of certain later developments of Bell's pioneering work, particularly of some quite recent ones that are not well known.

I also want to mention a non-historical event. From time to time I used to imagine an anniversary celebration of Bell's Theorem – the thirtieth anniversary

[1] The work for this paper was supported in part by the National Science Foundation, grant no. PHY9022345.

in 1994 being appropriate. Thirty years is long enough for the implications of the original work to be assessed with some perspective. The obvious locus for this celebration would have been at CERN. John Bell himself would have given his own survey of the consequences of his work and would have participated in the discussion of surveys presented by other workers in the field. It is very painful that this day-dream cannot be fulfilled, and that the present memorial volume must take the place of a celebration in which we could listen to John Bell himself.

2. The arguments of EPR and Bell

In 1935 A. Einstein, B. Podolsky, and N. Rosen (abbreviated EPR) [3] presented an argument that the quantum mechanical description of a physical system cannot be complete for all systems, specifically not for a pair of particles whose positions and linear momenta are both strictly correlated in a certain quantum state. I shall recapitulate EPR's argument, but shall follow D. Bohm [4] – as does Bell [1] – by adapting it to a system consisting of a pair of spin-1/2 particles in the quantum mechanical singlet spin state. This state can be represented (in Dirac's notation) by

$$|\psi\rangle = \frac{1}{\sqrt{2}}[|\hat{\mathbf{n}}+\rangle_1|\hat{\mathbf{n}}-\rangle_2 - |\hat{\mathbf{n}}-\rangle_1|\hat{\mathbf{n}}+\rangle_2], \tag{1}$$

where $\hat{\mathbf{n}}$ is a unit vector in Euclidean 3-space, $|\hat{\mathbf{n}}\pm\rangle_1$ represents a state of spin up (down) along $\hat{\mathbf{n}}$ for particle 1, and $|\hat{\mathbf{n}}\pm\rangle_2$ likewise for particle 2. The arbitrariness of the choice of $\hat{\mathbf{n}}$ in Eq. (1) indicates the spherical symmetry of the singlet spin state. There are two crucial philosophical assumptions in EPR's arguments, which I shall give in their own words:

i. (Reality Assumption): 'If, without in any way disturbing a system, we can predict with certainty (i.e., with probability equal to unity) the value of a physical quantity, then there exists an element of physical reality corresponding to this physical quantity.'
ii. (Locality Assumption): 'Since at the time of measurement the two systems no longer interact, no real change can take place in the second system in consequence of anything that may be done to the first system.'

The (adapted) argument of EPR proceeds as follows. If the two particles referred to in Eq. (1) are spatially well separated, then it is reasonable to assert the antecedent of the Locality Assumption ii, and hence the consequent. But by Eq. (1), a measurement of the spin of particle 1 along $\hat{\mathbf{n}}$, with result up or down, permits the inference with certainty that if spin along $\hat{\mathbf{n}}$ of particle 2 is measured the outcome will be respectively down or up. But the consequent of Assumption ii guarantees that particle 2 has in no way been disturbed by the measurement performed on particle 1. Consequently, by the Reality Assumption i, there is an element of physical reality corresponding to spin along $\hat{\mathbf{n}}$ of particle 2. But since the direction $\hat{\mathbf{n}}$ in Eq. (1) is arbitrary, precisely the same argument goes through for any direction. It follows that the spin along each direction $\hat{\mathbf{n}}$ corresponds to an element of physical reality of particle 2. The conclusion is that the quantum mechanical description of the system is incomplete, since neither the quantum state of Eq. (1) nor any quantum state can simultaneously specify the spin of a spin-1/2 particle in all directions – indeed, not in two non-parallel directions.

Bell [1] recapitulates this argument and introduces some notation for the elements of reality exhibited thereby. The element of reality corresponding to certainty of an outcome up (down) of the measurement of spin along $\hat{\mathbf{n}}$ of particle 1 is denoted by $A(\hat{\mathbf{n}}, \lambda) = \pm 1$, and $B(\hat{\mathbf{n}}, \lambda) = \pm 1$ has an analogous meaning for particle 2. Here λ denotes the complete state of the composite system $1 + 2$, in other words, the totality of its elements of reality. Since the quantum state $|\psi\rangle$ has been assumed to characterize the system under consideration correctly (even though incompletely), all λ consistent with this characterization are such that

$$A(\hat{\mathbf{n}}, \lambda) = -B(\hat{\mathbf{n}}, \lambda) \tag{2}$$

for each $\hat{\mathbf{n}}$. (In a more rigorous exposition, one could construe the expression 'with probability equal to unity' in the Reality Assumption as requiring Eq. (2) for all λ in a space Λ of complete states except for a set of measure zero, but in the present informal exposition nothing would be gained by this refinement.) Bell also assumes that – in the spirit of EPR, though going beyond their actual words – that in any experimental situation, such as the one in which an ensemble of pairs of spin-1/2 particles is prepared in the quantum state $|\psi\rangle$ – there is given a probability measure ρ over the space Λ of complete states.

Finally, Bell makes a locality assumption which is also in the spirit of EPR's Locality Assumption ii but quite different in formulation: that the expectation value of the product of outcomes of spin measurements on particles 1 and 2 of spin in the $\hat{\mathbf{a}}$ and $\hat{\mathbf{b}}$ directions, respectively, is given by

$$E_\rho(\hat{\mathbf{a}}, \hat{\mathbf{b}}) = \int_\Lambda A(\hat{\mathbf{a}}, \lambda) B(\hat{\mathbf{b}}, \lambda) \, \mathrm{d}\rho. \tag{3}$$

The crucial feature of Eq. (3) is the form of the integrand of the right hand side, for Bell writes, 'The vital assumption is that the result B for particle 2 does not depend on the setting $\hat{\mathbf{a}}$ of the magnet for particle 1, nor A on $\hat{\mathbf{b}}$.'

From Eqs. (2) and (3) Bell derives the inequality

$$1 + E_\rho(\hat{\mathbf{b}}, \hat{\mathbf{c}}) \geq |E_\rho(\hat{\mathbf{a}}, \hat{\mathbf{b}}) - E_\rho(\hat{\mathbf{a}}, \hat{\mathbf{c}})|. \tag{4}$$

This is the first of a family of inequalities involving correlations of pairs of systems, which have been derived since 1964 and are collectively known as 'Bell's Inequality.' In itself, Ineq. (4) is not striking except for its generality – it holds for any space Λ of complete states and for any distribution ρ, provided that Eqs. (2) and (3) are satisfied. The important thing is that there exist choices of directions $\hat{\mathbf{a}}$, $\hat{\mathbf{b}}$ and $\hat{\mathbf{c}}$ such that the quantum mechanical expectation values $E_\psi(\hat{\mathbf{a}}, \hat{\mathbf{b}})$, $E_\psi(\hat{\mathbf{a}}, \hat{\mathbf{c}})$, and $E_\psi(\hat{\mathbf{b}}, \hat{\mathbf{c}})$ do not satisfy Ineq. (4), where

$$E_\psi(\hat{\mathbf{a}}, \hat{\mathbf{b}}) = \langle \psi | (\boldsymbol{\sigma} \cdot \hat{\mathbf{a}})_1 (\boldsymbol{\sigma} \cdot \hat{\mathbf{b}})_2 | \psi \rangle, \tag{5}$$

etc. Thus there is a discrepancy between certain of the statistical predictions of quantum mechanics and any of the theories envisaged by EPR as supplementing the quantum mechanical description of the system under consideration. All such theories can be called 'local, deterministic hidden variables theories'. Here 'hidden variables' are those elements of a complete state λ, which are not contained in the quantum state; 'deterministic' means that the observable physical quantities are all assigned definite values when the complete state λ is specified; and 'local' strictly means that Eq. (3) is satisfied (though, as noted, this equation is in the spirit of EPR Locality Assumption ii). The discrepancy which Bell [1] demonstrates is now called 'Bell's Theorem', or, more accurately, it is a special case of the more general theorem that holds when the qualification 'deterministic' is dropped. Since the assumptions of Bell's own derivation are essentially those of EPR, with some supplementation that is entirely natural from their point of view, Bell has shown the impossibility of their vision regarding quantum mechanics: that is, to regard the quantum state as an incomplete characterization of an individual system but a correct statistical description of ensembles of similarly prepared systems. Had their vision been vindicated, it would have been possible to preserve all the statistical predictions of quantum mechanics but to eliminate the radical 'metaphysical' features of the orthodox interpretation of quantum mechanics, including objective indefiniteness of some physical quantities in every quantum state.

Bell's pioneering paper suggests a number of questions, which will be considered in the remainder of this chapter.

3. Is quantum mechanics required for deriving Bell's Inequality?

Bell's proof of Ineq. (4) uses the quantum mechanical state of Eq. (1) in two ways: first to derive the existence of the bivalent functions $A(\hat{\mathbf{n}}, \lambda)$ and $B(\hat{\mathbf{n}}, \lambda)$, and second to infer that for the same arguments these two functions have values of opposite sign, Eq. (2). Hence his argument is hybrid, combining assumptions of quantum mechanics and of local hidden variables theories. There is no logical weakness in this hybrid character, since essentially his theorem is the demonstration of an inconsistency of this combination.[2] But for two reasons one wishes a Bell's Inequality which does not make use of quantum mechanics: first, there is the intellectual curiosity to find out what is implicit in the hidden variables point of view by itself; and second, for the purpose of an experimental test of the point of view of local hidden variables theories it is desirable to have an experimental prediction which is derived without a quantum mechanical auxiliary assumption.

The desired demonstration was given by J. Clauser, M. Horne, A. Shimony and R. Holt [5], beginning with the assumption of a pair of systems, prepared in a complete state λ, each of which enters a two-channel analyzer with outcomes labeled $+1$ and -1. The analyzer of system 1 has variable parameter a, and that

[2] In the situation studied by Bell [1], local hidden variables theories are consistent with quantum mechanical perfect correlations (see the model of his Sect. 3) but inconsistent with quantum mechanical statistical correlations. By contrast, in the situations studied by Greenberger, Horne, and Zeilinger [48] and others, presented in Sect. 9, an inconsistency is exhibited already at the level of quantum mechanical perfect correlations.

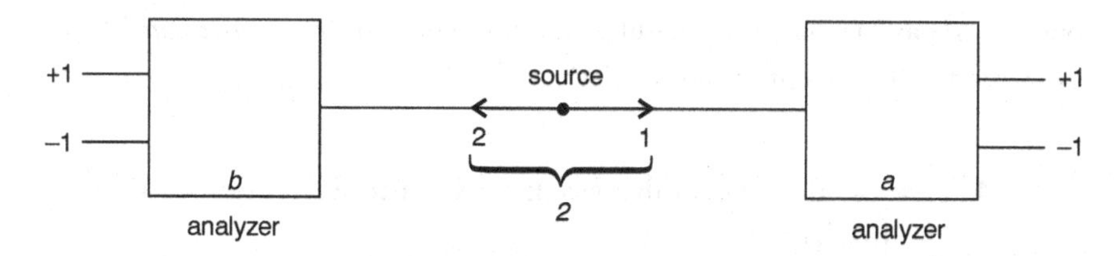

FIGURE 6.1 A particle pair is in complete state λ when it emerges from a source. Particle 1 is analyzed by the right analyzer, with a controllable parameter a and with two outcomes, conventionally assigned numerical values $+1$ and -1. Particle 2 is analyzed by the left analyzer, with a controllable parameter b and two outcomes assigned values $+1$ and -1.

of system 2 has a variable parameter b (see Fig. 6.1).[3] It is assumed that the outcome of the analysis of system 1 is a definite function of λ and the variable parameter a, expressed as $A(a, \lambda)$, and that the outcome of analysis of particle 2 is likewise a function $B(b, \lambda)$, so that

$$A(a, \lambda) = \pm 1, \quad B(b, \lambda) = \pm 1 \tag{6}$$

for all $\lambda \in \Lambda$, but no relation is assumed between the functions A and B. (It should be emphasized that the price of abstaining from a quantum mechanical assumption is that the definiteness of the functions A and B must be *postulated*, rather than *derived* as in the argument of EPR, which Bell was following.) Now, from Eqs. (3) and (6) one can easily derive the following inequality:

$$-2 \leq E_\rho(a', b') + E_\rho(a', b'') + E_\rho(a'', b') - E_\rho(a'', b'') \leq 2. \tag{7}$$

Here ρ is *any* distribution over Λ, rather than a distribution agreeing with some constraints derived from quantum mechanics, like Eq. (2). Bell's Theorem can be demonstrated via Ineq. (7), since there are abundant two-part systems, with appropriate settings of the parameters a and b, for which the quantum mechanical predictions violate this inequality.

Since quantum mechanics is not needed for the derivation of Ineq. (7), it could have been obtained by any classical probability theorist concerned with correlations and able to use probability distributions on continua, even

[3] Note that a and b are not restricted to be unit vectors in three-dimensional space, as in Bell [1], but are any variable parameters appropriate to the analyzers. In experiments so far the most important parameters have been the angle of orientation of a linear polarization analyzer and the phase angle induced by an optical phase shifter.

one as early as Jacob Bernoulli. But would there have been a *motivation* for the derivation before quantum mechanics?

4. Are bivalent variables essential for the derivation of Bell's Inequality?

No. If the range of $A(\hat{\mathbf{n}}, \lambda)$ is a discrete set of points in $[-1, 1]$, and likewise for the range of $B(\hat{\mathbf{n}}, \lambda)$, then an argument like that of Bell [6] will yield the same Ineq. (7). (Conceptually, this generalization is trivial, but practically it is very important, for it underlies an experiment in which each analyzer has three outcomes: detection in an upper channel, detection in a lower channel, and non-detection, which can be conveniently assigned the respective values $+1$, -1, and 0. The desirability of such an experiment will be discussed in Section 8 below.)

5. Is determinism essential for the derivation of Bell's Inequality?

A negative answer to this question was given by Bell [6], where footnote 10 should be especially noted) and by Clauser and Horne [7]. For their arguments it is of course essential to generalize the formulation of locality. Bell assumes that in an experiment on a pair of spin-1/2 particles the complete state λ and the direction $\hat{\mathbf{a}}$ determine not the outcome $A(\hat{\mathbf{a}}, \lambda)$ but rather its expectation value $\bar{A}(\hat{\mathbf{a}}, \lambda)$, the spread of results being due either to the hidden variables of the apparatus or to intrinsic indeterminism; and likewise λ and $\hat{\mathbf{b}}$ determine the expectation value $\bar{B}(\hat{\mathbf{b}}, \lambda)$. Both $\bar{A}(\hat{\mathbf{a}}, \lambda)$ and $\bar{B}(\hat{\mathbf{b}}, \lambda)$ lie in the closed interval $[-1, 1]$. The locality condition is that 'the values of $\bar{A}$ and $\bar{B}$ will be independent of $\hat{\mathbf{b}}$ and $\hat{\mathbf{a}}$ respectively'. The expression for the expectation value $E_\rho(a, b)$ of Eq. (3) is replaced by

$$E_\rho(\hat{\mathbf{a}}, \hat{\mathbf{b}}) = \int_\Lambda \bar{A}(\hat{\mathbf{a}}, \lambda)\bar{B}(\hat{\mathbf{b}}, \lambda)\, \mathrm{d}\rho \tag{8}$$

(where some modification of notation has been made), and then a new argument is given to derive Ineq. (7).

Clauser and Horne [7] work with probabilities rather than expectation values. With some modification of their notation the joint and single probabilities in the general situation described in Section 3 can be expressed as follows:

$p_\lambda(m, n|a, b)$ is the probability, if the complete state is λ and the parameters of analyzers 1 and 2 are respectively a and b, that the mth possible outcome will occur for the analysis of system 1 and the nth for the analysis of system 2.

$p^1_\lambda(m|a)$ is the probability that, if the complete state is λ and the parameter of analyzer 1 is a, the mth possible outcome will occur in the analysis of system 1.

$p^2_\lambda(n|b)$ is the probability that, if the complete state is λ and the parameter of analyzer 2 is b, the nth possible outcome will occur in the analysis of system 2.

Clauser's and Horne's adaptation of Bell's Locality Assumption is

$$p_\lambda(m, n|a, b) = p^1_\lambda(m|a)p^2_\lambda(n|b). \tag{9}$$

They derived a Bell's Inequality formulated explicitly in terms of joint and single probabilities in an ensemble, with distribution ρ over Λ:

$$\begin{aligned} -1 \le p_\rho(m, n|a', b') + p_\rho(m, n|a', b'') + p_\rho(m, n|a'', b') \\ - p_\rho(m, n|a'', b'') - p^1_\rho(m|a') - p_\rho(n|b') \le 0. \end{aligned} \tag{10}$$

Here

$$\begin{aligned} p_\rho(m, n|a, b) &= \int_\Lambda p_\lambda(m, n|a, b)\,\mathrm{d}\rho, \\ p^1_\rho(m|a) &= \int_\Lambda p^1_\lambda(m|a)\,\mathrm{d}\rho, \\ p^2_\rho(n|b) &= \int_\Lambda p^2_\lambda(n|b)\,\mathrm{d}\rho. \end{aligned} \tag{11}$$

Bell's Theorem can be demonstrated directly by exhibiting a quantum mechanical state and choices of parameters for which the quantum mechanical probabilities violate Ineq. (10); but alternatively the Theorem is demonstrated by deriving Ineq. (7) as a corollary of Ineq. (11) and using the quantum mechanical state that provides a violation of Ineq. (7).

The hidden variables theories considered by Bell [6] and by Clauser and Horne [7] are called 'stochastic', since the complete state λ, together with the values of the parameters, determines only expectation values or probabilities of outcomes of analysis, not the definite values of outcomes. Of course, if all probabilities are 1 or 0, the stochastic hidden variables theory reduces to a deterministic hidden variables theory as a limiting case.

6. Can Bell's Locality Assumption be analyzed further and understood more deeply?

The answer to this question is positive, though a mystery will be seen to persist. Much light was thrown on this question by a simple theorem of J. Jarrett [8].[4] He introduced two concepts, each of which catches a part of Bell's concept of locality. The first I call 'Parameter Independence' (a name which seems to me more descriptive of the meaning than Jarrett's name 'locality', which he must distinguish from Bell's concept by calling the latter 'strong locality'):

$$\text{Parameter Independence:} \quad p^1_\lambda(m|a,b) = p^1_\lambda(m|a), \qquad p^2_\lambda(n|a,b) = p^2_\lambda(n|b). \tag{12}$$

The second I call 'Outcome Independence' (which I think is more descriptive than Jarrett's 'completeness').

$$\text{Outcome Independence:} \quad p^1_\lambda(m|a,b,n) = p^1_\lambda(m|a,b), \qquad p^2_\lambda(n|a,b,m) = p^2_\lambda(n|a,b). \tag{13}$$

(The notation is almost self-explanatory: $p^1_\lambda(m|a,b,n)$ means the conditional probability of the mth outcome in an analysis of system 1 when λ, a and b are given, and in addition it is given that the nth possible outcome has occurred in an analysis of particle 2, etc.) Jarrett's theorem is that the conjunction of Parameter Independence and Outcome Independence is equivalent to Bell's Locality Assumption (in the form of Eq. (9)).

The significance of Jarrett's theorem is that the two components which he has distinguished have quite different relations to relativistic locality. To see this point clearly, consider an experimental arrangement like that of A. Aspect,

[4] Other theorists prior to Jarrett realized, with varying degrees of clarity, that Bell's concept of locality had two components. For instance, I heard this idea expressed in a lecture by C. Piron in 1981. But Jarrett introduced notation and terminology for the separate components and investigated explicitly the consequences of violating each of them separately.

J. Dalibard and G. Roger [9], in which the physical selection of the values of the parameters a and b are events of space-like separation. (Details are given in Aspect's paper in this volume.) The data of this experiment violate Bell's Inequality (7) and agree with the predictions of quantum mechanics. Since the inequality is a consequence of Bell's Locality Assumption, their result disconfirms the Assumption; and hence, by Jarrett's Theorem, the result disconfirms either Parameter Independence or Outcome Independence (or both). If the former, then the selection of parameter b affects the probabilities of outcomes of particle 1, or else the selection of parameter a affects the probabilities of outcomes of particle 2, in spite of space-like separation. If the latter, then the outcome of analysis of one particle affects the probabilities of the outcome of analysis of the other in a manner which is not 'programmed' in the complete state λ at the time of separation of the particles, again in spite of the space-like separation of the two analyses. Either alternative seems to require some kind of causal connection between events with space-like separation, contrary to relativity theory. The difference between the two alternatives is that violation of Parameter Independence permits in principle superluminal communication, whereas violation of Outcome Independence does not.

Violation of Parameter Independence: Consider a large ensemble of pairs of particles, all prepared in the complete state λ, and suppose that $p^2_\lambda(n|a', b) \neq p^2_\lambda(n|a'', b)$ for some n, b, a' and a'', and finally suppose that the switching mechanism of Aspect *et al.* is used to choose between a' and a'' rapidly enough to guarantee space-like separation from the analysis of particle 2 for each of the pairs of the ensemble. Let the experimenter located among the particles 1 of the ensemble make the same choice between a' and a'' in all cases. Then, if the ensemble is large enough, the statistics of analysis of the particles 2 in the ensemble permit an experimenter located among those particles to infer with arbitrarily high probability whether a' or a'' was selected – and thus a bit of information has been communicated superluminally. (P. Eberhard has pointed out privately that hidden variable theorists usually believe that experimentalists do not have sufficient control to prepare a system in a specific complete state λ, and therefore the superluminal communication capitalizing upon a violation of Parameter Independence would be possible only if it were possible to prepare an ensemble with distribution ρ over the space Λ of complete states in which the violation

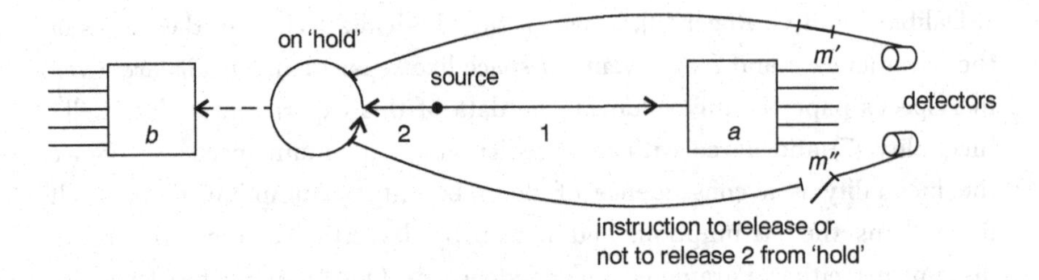

FIGURE 6.2 Particle 2 is placed on 'hold' until the analysis of particle 1 is completed. Only those particles 2 whose partners, particles 1, are detected in the channel m' (or m'', according to the experimenter's choice) are released from 'hold' and allowed to propagate to the analyzer on the left.

is not 'washed out', i.e.,

$$\int_{\Lambda} p_{\lambda}^{2}(n|a', b)\,\mathrm{d}\rho \neq \int_{\Lambda} p_{\lambda}^{2}(n|a'', b)\,\mathrm{d}\rho.) \tag{14}$$

If Outcome Independence is violated, there is an idealized technique which capitalizes upon the violation to send a message, as indicated in Fig. 6.2, but it will be obvious that the communication is not superluminal. Suppose that for outcomes m' and m'' of the analysis of particle 1 there is an inequality of conditional probabilities of outcome n in the analysis of particle 2: $p_{\lambda}^{2}(n|a, b, m') \neq p_{\lambda}^{2}(n|a, b, m'')$. Let a large ensemble of pairs be prepared in the complete state λ and let the particles 1 of each pair be analyzed, and the results of analysis be monitored by detectors. The particles 2 are placed on 'hold' (e.g., kept in a circular light guide) until instructions are transmitted from the locus of monitoring. The experimenter located there will make a decision after monitoring to release all particles 2 whose partners have outcomes m' and not release any others, or else to release just those particles 2 whose partners have outcomes m''. If the ensemble is large enough, an observer of the analyses of particles 2 can infer with arbitrarily high probability which choice was made by the first experimenter – and hence a bit of information is transmitted. But the time necessary for the complex process consisting of analyzing and monitoring particle 1, transmitting instructions to the place where particle 2 is on 'hold', and letting particle 2 propagate to its analyzer, will obviously take longer than a direct radar signal from the first to the second analyzer. Hence, the message is not transmitted superluminally.

Now the crucial question arises: since some predictions of quantum mechanics violate Bell's Inequality, as shown by Bell's Theorem, quantum mechanics must violate either Parameter Independence or Outcome Independence (or possibly both), but which? Part of the answer is trivial. Direct inspection of the quantum state $|\psi\rangle$ of Eq. (1) shows a violation of Outcome Independence: the probability of spin up in the direction $\hat{\mathbf{n}}$ for particle 2 is unity or zero according as the result of measurement of spin along $\hat{\mathbf{n}}$ for particle 1 is down or up, $|\psi\rangle$ being taken as the complete state of the particle. It is less obvious that quantum mechanics does not violate Parameter Independence, but this proposition has been demonstrated independently by Eberhard [10] Ghirardi–Rimini–Weber [11], and Page [12]. I shall not recapitulate any of their proofs, but shall only remark that the proofs depend essentially upon the linearity of quantum dynamics. It may be suspected, therefore, that if one postulates a nonlinear modification of quantum dynamics, such as that of S. Weinberg [13], one might encounter a violation of Parameter Independence. And indeed N. Gisin [14, 15] showed that Weinberg's theory has this peculiarity.

To summarize: quantum mechanics is in one sense (that is, in violating Outcome Independence) a nonlocal theory, and in so far as experiments confirm quantum mechanics against the family of local hidden variables theories, nature itself is nonlocal in the same way. Thus there is tension between quantum mechanics and relativity theory. But the impossibility of capitalizing upon a violation of Outcome Independence to signal superluminally implies a kind of 'peaceful coexistence' between quantum mechanics and relativity theory. In my opinion there is something still not understood about this 'peaceful coexistence'. We do not know what its implications are for space–time structure, for the concept of causality, and for the concept of an event. A number of remarks in Bell's papers (e.g., [2], pp. 60–1) and in his unpublished discussions show that he meditated deeply upon this mysterious problem, without finding a solution that satisfied him.

7. For what class of quantum states is there a violation of Bell's Inequality?

The singlet state of a pair of spin-1/2 particles, $|\psi\rangle$ of Eq. (1), has the peculiarity that it cannot be factorized in any way into a product of a quantum state of particle 1 and another quantum state of particle 2. In the locution of E. Schrödinger [16] $|\psi\rangle$ is 'entangled'. Likewise, the quantum state of the photon pairs used in the experiment of Aspect *et al.* is entangled. It turns out not to be accidental that the quantum states which exhibit a discrepancy with Bell's Inequality, and hence with the assumptions of local hidden variables theories, are entangled.

It is trivial to show that entanglement is a necessary condition for a violation of Bell's Inequality. Suppose the two-particle state $|\Phi\rangle$ can be written in product form as

$$|\Phi\rangle = |\phi\rangle_1 |\chi\rangle_2. \tag{15}$$

Let $P^1(a, m)$ be the projection operator on the Hilbert space $\mathscr{H}_1$ (associated with particle 1), whose eigenstates with eigenvalue 1 all ensure with certainty the mth outcome when a is the parameter of the analyzer of the particle; and $P^2(b, n)$ is an analogous projection operator on the Hilbert space $\mathscr{H}_2$ (associated with particle 2). Then the quantum mechanical expression for the probability of the joint results m and n satisfies the following:

$$\begin{aligned} p_{|\Phi\rangle}(m, n|a, b) &= \langle\Phi|P^1(a, m)P^2(b, n)|\Phi\rangle \\ &= \langle\phi|P^1(a, m)|\phi\rangle\langle\chi|P^2(b, n)|\chi\rangle. \end{aligned} \tag{16}$$

Hence the quantum mechanical joint probability satisfies the Locality Assumption of Eq. (9), from which Bell's Inequality follows. I believe that experts on Bell's Theorem were tacitly aware of this result for many years, but to my knowledge it was first stated explicitly by G.C. Ghirardi [17] and then by R. Werner [18].

Not trivial is the fact that the entanglement of a quantum state of a pair of spin-1/2 particles is a sufficient condition for violation of Bell's Inequality for an appropriate set of analyzers and parameters. This theorem is due to N. Gisin [19] and independently to Popescu and Rohrlich [20]. It seems to me straightforward to apply their argumentation to any pair of systems associated with Hilbert spaces of dimensions greater than or equal to two.

The question of this section can also be construed as inquiring about feasible experimental situations in which Bell's Inequality would be violated if the quantum mechanical predictions are fulfilled. The situation that Bell studied in his pioneering paper of 1964 [1] was realized once in an experiment analyzing S-wave scattering of a proton by a proton (Lamehi-Rachti and Mittig [21]), but it required several questionable auxiliary assumptions. Most of the successful experiments, including the pioneering experiment of Freedman and Clauser [22] and the one already noted of Aspect *et al.* [9], followed the suggestion of Clauser *et al.* [5] to study pairs of photons produced in an atomic cascade. Kasday, Ullman and Wu [23] used photon pairs produced in positronium annihilation and analyzed their polarizations by means of Compton scattering. All of the experiments mentioned so far analyzed the polarizations of a pair of particles (photons or protons). Bell [6] suggested an entirely different kind of test, using pairs of neutral kaons which could manifest themselves either as K^0 or as $\bar{K}^0$, but he expresses (footnote 13) doubts about the feasibility of such an experiment: 'Note that the spontaneous decay times of the two kaons, because they cannot be set at the will of the experimenter, are not to be regarded as analogous to the setting a and b of the Stern–Gerlach magnets. The thicknesses of a pair of slabs of matter placed in the lines of flight would be more relevant. I am told by Prof. B. d'Espagnat that the rapid decay of the short-lived kaon is a major obstacle to devising a crucial experiment.' A critical survey of attempts to test Bell's Inequality with kaons was recently given by Ghirardi, Grassi and Weber [24].

In recent years new experimental tests of Bell's Inequality with pairs of photons have been proposed and in one case executed, which differ from preceding experiments in that the entangled states involve linear momenta or directions rather than polarization. The first such proposal was made by Horne and Zeilinger [25], who considered a pair of photons produced by positronium at rest and impinging on crystal slabs M_1, M_2 (Fig. 6.3). The two-photon wave function is spherically symmetrical, but if one disregards all of it except the part that is Bragg diffracted by the crystals, the effective wave function is

$$|\psi\rangle = \frac{1}{\sqrt{2}}[|\mathbf{K}\rangle_1|-\mathbf{K}\rangle_2 + |\mathbf{K}'\rangle_1|-\mathbf{K}'\rangle_2]. \tag{17}$$

The diffracted beams are recombined at crystal slabs S_1 and S_2 after encountering plates which shift their phases by amounts θ_1 and θ_2 respectively. The recombined beams at S_1 feed two detectors R_1 and L_1, and the recombined beams at S_2 feed detectors R_2 and L_2. Thus the general scheme discussed above in Section 3 and depicted in Fig. 6.1 is exemplified, with R_1 and L_1 being the two possible outcomes of the analyzer of the first photon and θ_1 the

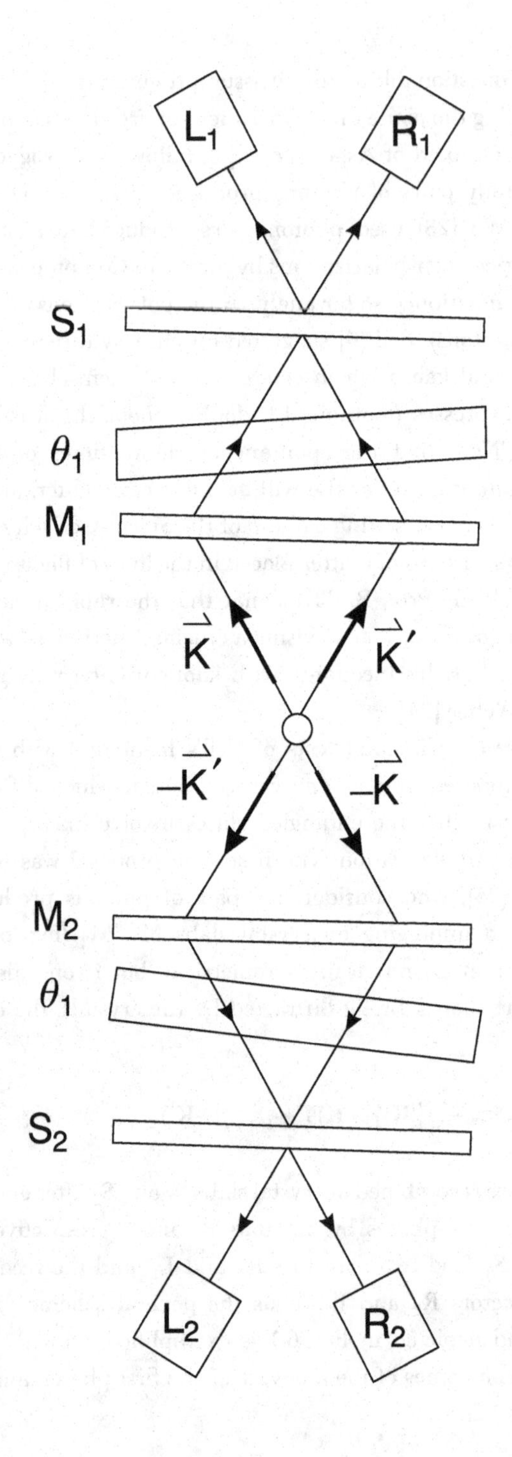
L1
R1
S1
θ1
M1
K
K′
−K′
−K
M2
θ1
S2
L2
R2

variable parameter of this analyzer, and likewise for R_2, L_2, θ_2. Unfortunately, the experimental realization of this proposal requires cooling the source well below a kelvin, because the extremely small width of the Bragg reflection curve in perfect crystal interferometers implies that both of a pair of correlated photons will be detected only if there is a stringent limit on thermal motion of the positronium. The motivation for this difficult experiment was diminished by the realization that pairs of photons produced by parametric down-conversion can exhibit the desired entanglement in momenta or directions.

The down-conversion process, in which a single photon incident upon a nonlinear crystal gives rise to a pair of correlated photons, was discovered by D.C. Burnham and D.L. Weinberg [26], and the quantum properties of the resulting pairs were extensively investigated by B.R. Mollow [27], L. Mandel [28] and many others. Down-conversion pairs of photons were used by Y.H. Shih and C. Alley [29] and by Z.Y. Ou and L. Mandel [30] to test Bell's Inequality, but in both experiments quarter-wave plates were introduced into the beams for the purpose of transforming momentum correlation into polarization correlation. That these pairs can be used for non-polarization tests of Bell's Inequality was proposed by M.D. Reid and D.F. Walls [31], P. Grangier, M.J. Potasek and B. Yurke [32], M. Horne, A. Shimony and A. Zeilinger [33], J.D. Franson [34], and J.G. Rarity and P.R. Tapster [35]; and the last two authors reported the first realization of such an experiment [36].

Figure 6.4 (taken from Horne *et al.* [33]) exhibits one arrangement for a nonpolarization test of Bell's Inequality. The source S is a nonlinear crystal pumped by a laser beam directed horizontally, which generates (with a small cross section) photon pairs of definite total linear momentum. Properly placed pinholes in a screen in front of the crystal allows a pair to emerge either in beams A and C or in beams D and B, in the sense that if a photon is detected in A its partner is detected in C, and likewise for D and B. It is essential, however, not to describe the ensemble of pairs as a mixture of $|A\rangle_1|C\rangle_2$ and $|D\rangle_1|B\rangle_2$ pairs, but rather to recognize that each pair is in the

FIGURE 6.3 (opposite) Positronium decays at rest into a pair of photons with total linear momentum zero, though the directions are indefinite. Directions $\mathbf{K}$ and $\mathbf{K}'$ are such that after Bragg diffraction by the perfect crystal M_1 and again by the perfect crystal S_1, photon 1 will enter either detector R_1 or L_1; and θ_1 is a variable phase shifter. Likewise for

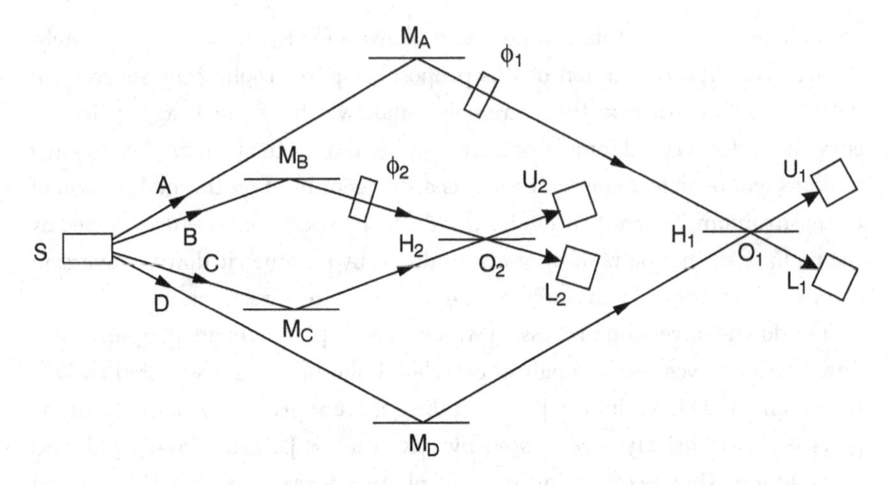

FIGURE 6.4 Two photons emerge from the source in the entangled state of Eq. (18). The two beams A and D of photon 1 are combined near point O_1 on the half-silvered mirror H_1 (Mach–Zehnder interferometer) and phase shifter ϕ_1 is placed in beam A. Likewise the two beams B and C of particle 2 are combined near point O_2 of half-silvered mirror H_2, and phase shifter ϕ_2 is placed in beam B. Coincidence counts at the four pairs of detectors – U_1U_2, U_1L_2, L_1U_2, L_1L_2 – are observed.

entangled state

$$|\psi\rangle = \frac{1}{\sqrt{2}}[|A\rangle_1|C\rangle_2 + |D\rangle_1|B\rangle_2], \tag{18}$$

as can be confirmed indirectly by experimental verifications of the probabilities of joint detection of the photons which will be derived, in a manner to be described, from Eq. (18). There are two paths for photon 1 en route to the half-silvered mirror H_1, namely A and D, and in the former a plate with a variable phase shift ϕ_1 is inserted. Likewise, there are two paths for photon 2 en route to the half-silvered mirror H_2, B and C, and in the former a plate with a variable phase shift ϕ_2 is inserted. H_1 feeds an upper detector U_1 and a lower detector L_1; and likewise H_2 feeds detectors U_2 and L_2. The detectors will be assumed for simplicity to be perfectly efficient, and H_1 and H_2 will be taken to be lossless symmetric beam-splitters, such that transmission is associated with a factor $2^{-1/2}$ and reflection with a factor $2^{-1/2}i$ (which is a special case of a more general relation between reflected and transmitted beams that leads to essentially the same expression for the probability of joint detection). Then the total probability amplitude for joint detection of the two photons in U_1 and U_2 is

$$A_\psi(U_1, U_2|\phi_1, \phi_2) = 2^{-1/2}[(2^{-1/2}i\, e^{i\phi_1})(2^{-1/2})\, e^{iw_1} + (2^{-1/2})(2^{-1/2}i\, e^{i\phi_2})\, e^{iw_2}], \tag{19}$$

where w_1 is a phase angle depending upon the detailed placement of the optical elements of the two-photon path A–C except for the variable phase shift ϕ_1, and w_2 is a phase angle depending upon the detailed placement of the optical elements of the two-photon path D–B except for ϕ_2. The absolute square of $A_\psi(U_1, U_2|\phi_1, \phi_2)$ is the probability of joint detection by the upper detectors:

$$p_\psi(U_1, U_2|\phi_1, \phi_2) = \tfrac{1}{4}[1 + \cos(\phi_2 - \phi_1 + w_2 - w_1)]. \tag{20}$$

Similarly

$$p_\psi(L_1, L_2|\phi_1, \phi_2) = \tfrac{1}{4}[1 + \cos(\phi_2 - \phi_1 + w_2 - w_1)], \tag{21}$$

and

$$p_\psi(U_1, L_2|\phi_1, \phi_2) = p_\psi(L_1, U_2|\phi_1, \phi_2) = \tfrac{1}{4}[1 - \cos(\phi_2 - \phi_1 + w_2 - w_1)]. \tag{22}$$

If the upper detectors are (conventionally) assigned the numerical value +1 and the lower detectors the numerical value −1, then the expectation value of the product of the numerical outcomes is

$$E_\psi(\phi_1, \phi_2) = \cos(\phi_2 - \phi_1 + w_2 - w_1). \tag{23}$$

We can now make a comparison with Bell's Inequality (in the version of Ineq. (7)), letting the variable parameters *a* and *b* become the variable phase shifts ϕ_1 and ϕ_2. Let the phase shifts have the following values:

$$\phi_1' = \tfrac{1}{2}\pi, \quad \phi_2' = \tfrac{1}{4}\pi + w_1 - w_2, \quad \phi_1'' = 0, \quad \phi_2'' = (3\pi/4) + w_1 - w_2. \tag{24}$$

Then

$$\begin{aligned}\cos(\phi_2 - \phi_1 + w_2 - w_1) &= \cos(\phi_2' - \phi_1'' + w_2 - w_1) \\ &= \cos(\phi_2'' - \phi_1' + w_2 - w_1) \\ &= -\cos(\phi_2'' - \phi_1'' + w_2 - w_1) = 0.707. \end{aligned} \tag{25}$$

Hence,

$$E_\psi(\phi_1', \phi_2') + E_\psi(\phi_1', \phi_2'') + E_\psi(\phi_1'', \phi_2'') - E_\psi(\phi_1'', \phi_2'') = 2.828, \qquad (26)$$

in violation of Ineq. (7).

The experiment of Rarity and Tapster [36] is similar in basic design to the idealized experiment just sketched, though it differs in many practical details. Their results agreed with the predictions of quantum mechanics and deviated by ten standard deviations from the extreme limit of Bell's Inequality.

8. How decisively is the family of local hidden variables theories experimentally refuted?

Of the thirteen tests of Bell's Inequality listed by M. Redhead [37], all but two (R. Holt and F. Pipkin, unpublished, and G. Faraci, S. Gutkowski, S. Notarrigo, and A.R. Pennisi [38]) agreed with the predictions of quantum mechanics and violated Bell's Inequality. Later tests, such as those mentioned in the preceding section, have strengthened the case against Bell's Inequality and hence against the assumptions of local hidden variables theories from which it follows. A discussion of possible systematic errors in the experiment of Holt and Pipkin is given by Clauser and Shimony [39], pp. 1910–11. Even if the experimental results were fairly evenly balanced *pro* and *con*, there would be a good methodological reason to give greater credit to those which favored quantum mechanics, entirely independent of considerations of the past successes of that theory. In the situations of the experiments, quantum mechanics predicts very strong correlations as a result of entanglement; whereas the Locality Assumption is a constraint upon correlations. Random or small systematic errors are more likely to wash out strong correlations, thereby speciously giving results agreeing with Bell's Inequality, than to produce coincidentally the strict correlations of quantum mechanics if nature were really governed by a local hidden variables theory.

Nevertheless, there are two great lacunae in most of the inferences from experimental results to the rejection of the family of local hidden variables theories. The first is the 'communication loophole': that the parameters of the two analyzers remain fixed for sufficiently long periods of time for the mechanism of analysis by analyzer 1 to take account of the parameter b of analyzer 2 and vice versa, and if so there could be a violation of Parameter Independence (Eq. 12) and hence of the Locality Assumption (Eq. 9) without any violation of relativistic space–time structure. The experiment of Aspect *et al.* [9], with rapidly varied polarization analyzers, was designed specifically to block this loophole, and the fact that it yielded results in excellent agreement with quantum mechanics and discordant with Bell's Inequality is generally regarded as a spectacular disconfirmation of local hidden variables theories. To be sure, the fact that the variation of the polarization analyzers is periodic opens the possibility that information about parameter values can still be transferred from one side of the experiment to the other without violating relativity theory, but most students of the subject doubt that nature would have utilized such a surreptitious device for misleading the experimenters.

The other is the 'selection loophole'. In all experiments so far the detectors have been far less than ideal, for example, about 20% quantum efficiency for the typical photo-detectors in the two-photon experiments, so that only about 4% of the pairs which jointly pass into two output channels are detected. This loophole can be blocked by making a fair sampling assumption (Clauser *et al.* [5]) that the proportion of pairs detected is independent of the choice of pairs of output channels and of the analyzer parameters. But there is no clear way to test the fair sampling assumption itself, and by denying it one can construct *local* models, in which the hidden variables govern not only the outcome of analysis but also detection or non-detection, such that the resulting counting rates agree with quantum mechanics and violate Bell's Inequality (Clauser and Horne [7], Marshall, Santos and Selleri [40]).

With sufficiently high, though less than ideal, efficiency the selection loophole can nevertheless in principle be blocked experimentally. Returning to the discussion of Section 4, one can consider analyzers with three output channels identified as follows: the first is detection by a detector placed in an upper position (e.g. in the path of the ordinary ray from a Wollaston prism), the second is detection by a detector placed in a lower position (e.g., in the path of the extraordinary ray from a Wollaston prism) and the third is non-detection. The values $+1$, -1, and 0 are associated with these three channels. Now if the Locality Assumption (Eq. 9) holds, or equivalently the conjunction of Parameter Independence and Outcome Independence, then Ineq. (7) follows; and because of the specifications of the three output channels the expectation values $E_\rho(a, b)$ are experimental quantities, provided that the number of pairs in the ensemble is determined by calorimetry or some other means. A simple calculation (Shimony [41]) shows that the quantum mechanical predictions violate Ineq. (7) if the efficiencies of the detectors are greater than 0.841. A less stringent requirement, namely that the efficiencies be greater than 0.828, is demonstrated by more complicated arguments of N.D. Mermin and G. Schwarz [42] and A. Garg and N.D. Mermin [43]. Evidently an experiment that meets this demand and thereby blocks the detection loophole is very difficult, but there is a prospect that such an experiment will be performed. Edward Fry, Thomas Walther and Shifang Li (Texas A&M University, unpublished) have designed a molecular experiment which will fulfill the requirements for efficiencies of detectors and also of analyzers, and which therefore is able to dispense with auxiliary assumptions such as fair sampling. Furthermore, the experiment will simultaneously block the communication loophole. The experiment is in progress.

9. Is it possible to demonstrate Bell's Theorem without Bell's Inequality?

A positive answer has been demonstrated in two entirely different ways. In the early 1970s S. Kochen considered a pair of spin-1 particles prepared in the singlet state. Each component of spin $s_{\mathbf{n}}(1)$ of particle 1 can have values of 1, 0, -1 in units of $\hbar$, and if it is determined by measurement to have one of these values then one can infer with certainty that $s_{\mathbf{n}}(2)$ has the negative of this value. A corollary is that if $s_{\mathbf{n}}^2(1)$ is found to have value 0 or 1, then it can be inferred with certainty that $s_{\mathbf{n}}^2(2)$ has the same value. Furthermore, $s_{\mathbf{n}_1}^2(1)$, $s_{\mathbf{n}_2}^2(1)$, and $s_{\mathbf{n}_3}^2(1)$ commute and hence can be simultaneously measured; and if so, their sum is necessarily 2. The values of $s_{\mathbf{n}_1}^2(2)$, $s_{\mathbf{n}_2}^2(2)$, and $s_{\mathbf{n}_3}^2(2)$ can then be inferred with certainty, and hence by Assumptions i and ii of EPR there is an element of physical reality corresponding to every quantity $s_{\mathbf{n}}^2(2)$ of particle 2 (which specifically is independent of what other directions are taken along with $\mathbf{n}$ to form an orthogonal triad), with values 1 or 0, and with the additional property that the sum of the squares of the spins in any three orthogonal directions is 2 (in units of $\hbar$). But it follows as a corollary of a theorem of A. Gleason [44], or alternatively from theorems of J.S. Bell [45] or of S. Kochen and E. Specker [46], that such a function of $\mathbf{n}$ into the bivalent set $\{1, 0\}$ is impossible. Hence there is a contradiction between the assumptions of EPR and the *non-statistical* predictions of the quantum mechanical singlet state of a pair of spin-1 systems. This argument differs from that of Bell [1] and all arguments inspired by Bell in that the contradiction only concerns the algebraic structure of the lattice of propositions of the spin-1 system and the perfect correlations of a suitably prepared pair of spin-1 systems, but never any imperfect or merely statistical correlations. Kochen never published this argument, but it was known by a few people in the oral tradition and was independently discovered by P. Heywood and M. Redhead [47]. It should be remarked that the argument is neither simple nor transparent, since it depends upon a choice among previous theorems, none of which is very simple.

The second argument, by contrast, is extraordinarily simple and transparent. It was invented by D. Greenberger, M. Horne, and A. Zeilinger [48] and was clarified in various ways by N.D. Mermin [49], Greenberger, Horne, Shimony, and Zeilinger [50], and Clifton, Redhead, and Butterfield [51]. The original presentation by Greenberger *et al.* [48] studied a properly prepared state of four spin-1/2 systems, but I shall present a version of Greenberger *et al.* [50] which makes use of three-photon interferometry. Figure 6.5 shows three photons generated by down-conversion from a single photon incident perpendicular to the paper; the propagation of the three daughter photons is projected into the plane of the paper, and the momentum along the incident direction is suppressed. There are three pairs of pinholes in a cylindrical shield about the source, with photon 1 going into beams a or a', photon 2 into beams b or b', and photon 3 into beams c or c', but of course the 'or' is to be interpreted non-classically, not implying that each photon makes a definite choice of one of the beams. In fact, the quantum mechanical state is assumed to be

$$|\psi\rangle = \frac{1}{\sqrt{2}}[|a\rangle_1|b\rangle_2|c\rangle_3 + |a'\rangle_1|b'\rangle_2|c'\rangle_3], \tag{27}$$

which is entangled. Beams a and a' are brought together at a beam splitter, which feeds two detectors d and d′; and a phase plate causing an adjustable phase shift ϕ_1 is inserted in the path a'. Similarly b and b' are brought together and then feed detectors e and e′, and a plate with phase shift ϕ_2 is placed in the path b'; and c and c' are brought together and feed detectors f and f′, and a plate with phase shift ϕ_3 is placed in path c'. The numerical value $+1$ is

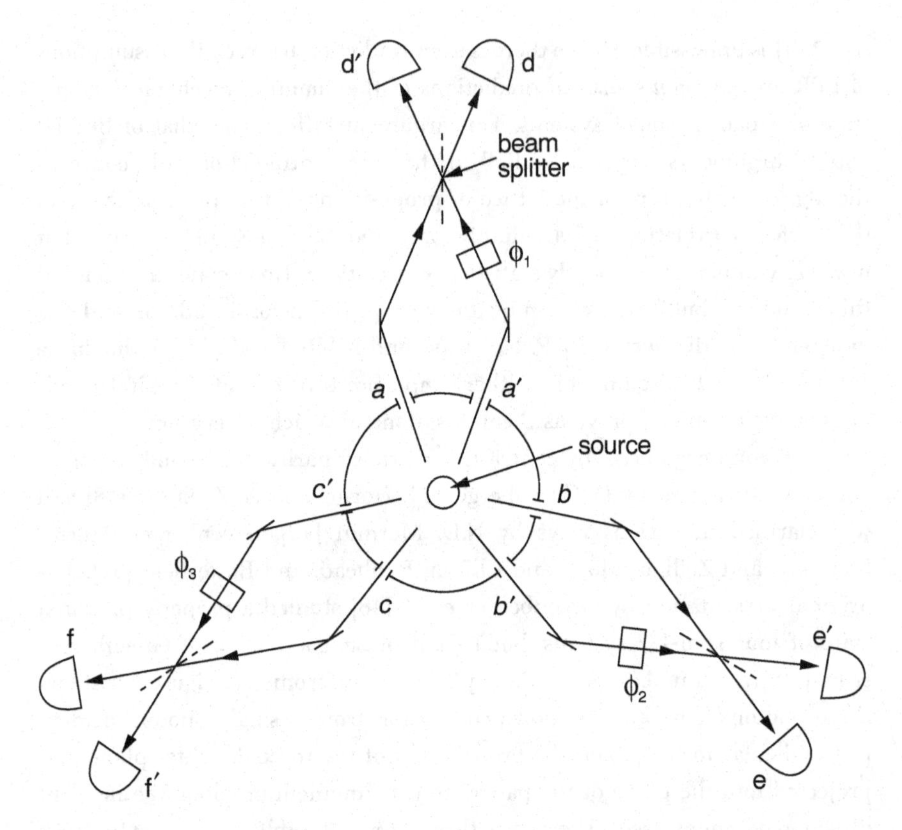

FIGURE 6.5 Three-photon interferometry. The three photons emerge from the slits in the entangled state of Eq. (26). Beams a and a' of photon 1 are brought together at a half-silvered mirror which feeds detectors d and d′, and a variable phase shifter ϕ_1 is placed in the beam a'. Likewise for photons 2 and 3.

associated with the unprimed detectors d, e, f, and the value -1 is associated with the primed detectors d′, e′, f′. A straightforward interferometric calculation like that sketched in Section 7 shows that the probability of triple outcomes x, y, z is

$$P_{\psi}(x,y,z|\phi_1,\phi_2,\phi_3) = \tfrac{1}{8}[1 \pm \sin(\phi_1 + \phi_2 + \phi_3)], \tag{28}$$

where x is d or d′, y is e or e′, z is f or f′, and the spin is $+$ or $-$ according as the number of primes among the x, y, z is even or odd. With the numerical values conventionally assigned to the detectors, the expectation value of the product of the outcomes is

$$E_{\psi}(\phi_1, \phi_2, \phi_3) = \sin(\phi_1 + \phi_2 + \phi_3). \tag{29}$$

We shall be particularly interested in two special cases:

$$E_{\psi}(\phi_1, \phi_2, \phi_3) = 1 \quad \text{if } \phi_1 + \phi_2 + \phi_3 = \pi/2 (\text{mod } 2\pi), \tag{30}$$

$$E_{\psi}(\phi_1, \phi_2, \phi_3) = -1 \quad \text{if } \phi_1 + \phi_2 + \phi_3 = 3\pi/2 (\text{mod } 2\pi). \tag{31}$$

Hence, if the phase shifts satisfy the conditions of either (30) or (31), one can predict with certainty which detector will be triggered by one of the photons if we are told which ones are triggered by the other two photons. But because of the spatial separation of the photons after they emerge from their respective pinholes we are able to apply EPR's Locality Assumption ii, and then use its consequent together with the Reality Assumption i to conclude that the procliv-ity to enter a specific one of its two possible detectors is an element of reality for each of the three photons. Hence for each complete state λ of the three photons compatible with the quantum state of Eq. (26) there is a function $A_{\lambda}(\phi_1)$ with values $+1$ or -1 according as photon 1 is destined to trigger d or d$'$; a function $B_{\lambda}(\phi_2) = 1$ or -1 as photon 2 is destined to trigger e or e$'$; and a function $C_{\lambda}(\phi_3) = 1$ or -1 as photon 3 is destined to trigger f or f$'$. Agreement with the quantum mechanical prediction (30) implies the following equations for each λ compatible with the quantum state of Eq. (27):

$$A_{\lambda}(\pi/2)B_{\lambda}(0)C_{\lambda}(0) = 1, \tag{32a}$$

$$A_{\lambda}(0)B_{\lambda}(\pi/2)C_{\lambda}(0) = 1, \tag{32b}$$

$$A_{\lambda}(0)B_{\lambda}(0)C_{\lambda}(\pi/2) = 1. \tag{32c}$$

Multiplying the right hand sides and the left hand sides of Eqs. (32a, b, c) yields

$$A_{\lambda}(\pi/2)B_{\lambda}(\pi/2)C_{\lambda}(\pi/2) = 1, \tag{33}$$

which conflicts with the quantum mechanical prediction of Eq. (31). (Note that this argument neglects the difference between 'true for all λ compatible with Eq. (27)' and 'true for all but a set of measure zero of such λ', but one obtains the same result even if rigorous measure theory is used.) An inconsistency is thus exhibited between the existence of the bivalent functions $A_{\lambda}(\phi_1)$, $B_{\lambda}(\phi_2)$, $C_{\lambda}(\phi_3)$ – which is the crucial content of a deterministic local hidden variables theory for the three-photon system – and the *non-statistical* quantum predictions from $|\psi\rangle$ of Eq. (28). This inconsistency is a version of Bell's

Theorem, which resembles the version of Bell [1] in that it concerns only deterministic hidden variables theories, but of course differs from it by using no inequalities and by restricting its attention to non-statistical quantum predictions. Since the existence of the bivalent functions $A_\lambda(\phi_1)$, $B_\lambda(\phi_2)$, $C_\lambda(\phi_3)$ follows from the conjunction of EPR's Assumptions i and ii with the non-statistical predictions of quantum state $|\psi\rangle$, the inconsistency of this conjunction has also been exhibited.

It may be noted, in case someone wishes to realize the three-photon state of Eq. (27) in a laboratory, that the generation of triplets of photons by down-conversion seems never to have been observed so far. It is possible theoretically to construct a four-photon state with correlations incompatible with any local deterministic hidden variables theory, and this state might be produced by generating a pair of photons via down-conversion by an initial pumping photon and letting each of these generate two others in two separate nonlinear crystals. Since the two latter nonlinear crystals could be pumped by auxiliary laser beams (as suggested in conversation by L. Mandel), the cross-section for the production of the desired quadruple of photons might be greatly enhanced. For the present this is only a theoretician's experiment. Other proposals for producing entangled n-photon states ($n \geq 3$) have been made by Horne and Zeilinger (unpublished) and by Rubin and Shih (unpublished).

10. Concluding remarks

The concentration in this chapter upon various aspects of Bell's Theorem should not mislead a reader into thinking that Bell's own work on foundations of quantum mechanics was equally specialized. Indeed, what troubled Bell most about the present formulations of quantum mechanics is not quantum non-locality, but rather the pervasive division of the physical world into that which is observed and that which is the tool of the observer. His preoccupation with this problem is frequently expressed in the wonderful collection of essays *Speakable and Unspeakable in Quantum Mechanics* [2], especially in essay 5, entitled 'Subject and object'. This book should be read in its entirety to see how his work on local hidden variables theories is interwoven with his deep concern with other aspects of quantum mechanics and of natural philosophy in general. Another beautiful presentation is 'Against "Measurement"' [52].

Appendix: A proof of Bell's Inequality

In order to make this chapter more self-contained a proof will be given of one version of Bell's Inequality, Ineq. (7).

LEMMA: If x', y', x'', y'' all belong to the interval $[-1, 1]$, and $S = x'y' + x'y'' + x''y' - x''y''$, then S lies in the interval $[-2, 2]$.

PROOF (suggested by N.D. Mermin): Since S is linear in all four of its variables, it assumes its extreme values at the corners of the hypercube, where (x', y', x'', y'') equals $(\pm 1, \pm 1, \pm 1, \pm 1)$. At the corners S clearly must be an integer between -4 and 4. But since S can be written as $(x' + x'')(y' + y'') - 2x''y''$, and the quantities in parentheses can only be 0 or ± 2, the values ± 3 and ± 4 at the corners are excluded.

Let s_m be the numerical value of the mth outcome of the analysis of particle 1 and t_n the numerical value of the nth outcome of the analysis of particle 2. The expectation value of the product of these outcomes is

$$\begin{aligned} E_\lambda(a, b) &= \sum_{m,n} s_m t_n p_\lambda(m, n|a, b) = \sum_{m,n} s_m t_n p_\lambda^1(m|a) p_\lambda^2(n|b) \\ &= \left(\sum_m s_m p_\lambda^1(m|a)\right)\left(\sum_n t_n p_\lambda^2(n|b)\right) \\ &= E_\lambda^1(a) E_\lambda^2(b), \end{aligned} \tag{A1}$$

where the Locality Assumption Eq. (9) has been used and the expectation values $E_\lambda(a, b)$, $E_\lambda^1(a)$ and $E_\lambda^2(b)$ have the obvious meanings given in context. If s_m and t_n are all in the interval $[-1, 1]$, as in Section 4, then x', y', x'', y'' of the lemma can be taken to be $E_\lambda^1(a')$, $E_\lambda^2(b')$, $E_\lambda^1(x'')$, $E_\lambda^2(b'')$ respectively, and hence from the lemma

$$-2 \leq E_\lambda(a', b') + E_\lambda(a', b'') + E_\lambda(a'', b') - E_\lambda(a'', b'') \leq 2. \tag{A2}$$

Integrating all terms of Ineq. (A2) with the probability distribution ρ yields Ineq. (7).

Note that this proof holds for non-deterministic hidden variables theories and for any set of discrete outcomes from the analyzers, as claimed in Sections 3 and 4.

Appendix

Since the main text of this chapter was submitted there were two independent proofs of an important theorem showing that any pure entangled quantum state (of two or more subsystems, associated with Hilbert spaces of any dimensionality) entails a violation of Bell's Inequality for some choice of observables.

The references are:

N. Gisin, *Phys. Lett. A* **145** (1991), 145
and
S. Popescu and D. Rohrlich, *Phys. Lett. A* **166** (1992), 293.

Further comments relating to this theorem are given in:

S. Popescu, in *The Dilemma of Einstein, Podolsky and Rosen – 60 Years After*, eds. A. Mann and M. Revzen (IOP Publishing, 1996).

REFERENCES

[1] J.S. Bell, *Physics* **1** (1964) 195 (reprinted in [2]).
[2] J.S. Bell, *Speakable and Unspeakable in Quantum Mechanics* (Cambridge University Press, Cambridge, U.K., 1987).
[3] A. Einstein, B. Podolsky and N. Rosen, *Phys. Rev.* **47** (1935) 777.
[4] D. Bohm, *Quantum Theory* (Prentice-Hall, Englewood Cliffs, N.J., 1951).
[5] J.F. Clauser, M.A. Horne, A. Shimony and R.A. Holt, *Phys. Rev. Lett.* **23** (1969) 880.
[6] J.S. Bell, in *Foundations of Quantum Mechanics*, ed. B. d'Espagnat (Academic Press, New York, 1971), 171 (reprinted in [2]).
[7] J.F. Clauser and M.A. Horne, *Phys. Rev.* **D10** (1974) 526.
[8] J. Jarrett, *Nous* **18** (1984) 569–89.
[9] A. Aspect, J. Dalibard and C. Roger, *Phys. Rev. Lett.* **49** (1982) 1804.
[10] P. Eberhard, *Nuovo Cimento* **38B** (1977) 75.
[11] G.C. Ghirardi, A. Rimini and T. Weber, *Nuovo Cimento Lett.* **27** (1980) 293.
[12] D. Page, *Phys. Lett.* **91A** (1982) 57.
[13] S. Weinberg, *Annals of Phys. (New York)* **194** (1989) 336.
[14] N. Gisin, *Helv. Phys. Acta* **62** (1989) 363.
[15] N. Gisin, *Phys. Lett.* **A143** (1990) 1.
[16] E. Schrödinger, *Cambridge Phil. Soc. Proc.* **31** (October 1935) 555–63.
[17] G.C. Ghirardi, in *Dynamical Systems and Microphysics*, eds. A. Blaquiere, F. Fer and A. Marzollo (Springer-Verlag, Wien – New York, 1980) 366.
[18] R.F. Werner, *Phys. Rev.* **A40** (1989) 4277.
[19] N. Gisin, *Phys. Lett.* **A154** (1991) 201.

[20] S. Popescu and D. Rohrlich, *Phys. Lett.* **A169** (1992) 411.

[21] M. Lamehi-Rachti and W. Mittig, *Phys. Rev.* **D14** (1976) 2543.

[22] S.J. Friedman and J.F. Clauser, *Phys. Rev. Lett.* **28** (1972) 938.

[23] L.R. Kasday, J.D. Ullman and C.S. Wu, *Nuovo Cimento* **25B** (1975) 633.

[24] G.C. Ghirardi, R. Grassi and T. Weber, *Quantum Mechanics Paradoxes at the Φ-Factory* (University of Trieste preprint, unpublished, 1991).

[25] M.A. Horne and A. Zeilinger, in *Proceedings of the Symposium on the Foundations of Modern Physics*, eds. P. Lahti and P. Mittelstaedt (World Scientific, Singapore, 1985) 435.

[26] D.C. Burnham and D.L. Weinberg, *Phys. Rev. Lett.* **25** (1970) 84.

[27] B.R. Mollow, *Phys. Rev.* **A8** (1973) 2684.

[28] L. Mandel, *Phys. Rev.* **A28** (1983) 929.

[29] Y.H. Shih and C.O. Alley, *Phys. Rev. Lett.* (USA) **61** (1988) 2921.

[30] Z.Y. Ou and L. Mandel, *Phys. Rev. Lett.* (USA) **61** (1988) 50.

[31] M.D. Reid and D.F. Walls, *Phys. Rev.* **A34** (1986) 1260.

[32] P. Grangier, M.J. Potasek and B. Yurke, *Phys. Rev.* **A38** (1988) 3132.

[33] M.A. Horne, A. Shimony and A. Zeilinger, *Phys. Rev. Lett.* **62** (1989) 2209.

[34] J.D. Franson, *Phys. Rev. Lett.* **62** (1989) 2205.

[35] J.G. Rarity and P.R. Tapster, *Phys. Rev.* **A41** (1990) 5139.

[36] J.G. Rarity and P.R. Tapster, *Phys. Rev. Lett.* **64** (1990) 2495.

[37] M.L.G. Redhead, *Incompleteness, Nonlocality, and Realism* (Oxford University Press, Oxford, 1987).

[38] G. Faraci, S. Gutkowski, S. Notarrigo and A.R. Pennisi, *Nuovo Cimento Lett.* **9** (1974) 607.

[39] J.F. Clauser and A. Shimony, *Rep. Progr. Phys.* **41** (1978) 881.

[40] T.W. Marshall, E. Santos and F. Selleri, *Phys. Lett.* **98A** (1983) 5.

[41] A. Shimony, in *Sixty-Two Years of Uncertainty: Historical, Philosophical, and Physical Inquiries into the Foundations of Quantum Mechanics*, ed. A. Miller (Plenum, New York, 1990) 33.

[42] N.D. Mermin and G.M. Schwarz, *Found. Phys.* **12** (1982) 101.

[43] A. Garg and N.D. Mermin, *Phys. Rev.* **D35** (1987) 3831.

[44] A. Gleason, *J. Math. Mech.* **6** (1957) 885.

[45] J.S. Bell, *Rev. Mod. Phys.* **38** (1966) 447.

[46] S. Kochen and E. Specker, *J. Math. Mech.* **17** (1967) 59.

[47] P. Heywood and M.L.G. Redhead, *Found. Phys.* **13** (1983) 481.

[48] D.M. Greenberger, M.A. Horne and A. Zeilinger, in *Bell's Theorem, Quantum Theory, and Conceptions of the Universe*, ed. M. Kafatos (Kluwer Academic, Dordrecht, 1989) 73.

[49] N.D. Mermin, *Am. J. Phys.* **58** (1990) 731.

[50] D.M. Greenberger, M.A. Horne, A. Shimony and A. Zeilinger, *Am. J. Phys.* **58** (1990) 1131.

[51] R.K. Clifton, M.L.G. Redhead and J.N. Butterfield, *Found. Phys.* **21** (1991) 149.

[52] J.S. Bell, in *Sixty-Two Years of Uncertainty: Historical, Philosophical and Physical Inquiries into the Foundations of Quantum Mechanics*, ed. A. Miller (Plenum, New York, 1990).

7 Does quantum mechanics carry the seeds of its own destruction?*

KURT GOTTFRIED

> ... the quantum mechanical description will be superseded. In this it is like all theories made by man. But to an unusual extent its ultimate fate is apparent in its internal structure. It carries within itself the seeds of its own destruction [1].

I

Quantum mechanics is the most powerful theory of physics. It was John Bell's view, however, that no formulation of orthodox quantum mechanics is free of fatal flaws. While fully recognizing the theory's indisputable utilitarian success, Bell was convinced that its failings meant that 'something is rotten in the state of Denmark'. Indeed, as many know from personal experience [2], he was in the habit of proclaiming that quantum mechanics is 'rotten'. What I do not know is whether he chose this particular adjective as a veiled condemnation of the House of Bohr.

What are the seeds of destruction that he saw so clearly?

To approach this question it is best to first examine what Bell meant by a satisfactory theory. I was fortunate to gain an insight into this through several long discussions we had during a bucolic June week in 1990 at a wonderful workshop on the foundations of quantum mechanics held at Amherst College. Bell held out classical electrodynamics as the paradigm of a properly formulated theory of a 'serious piece of physics'. Classical electrodynamics is, however, beset by serious internal difficulties when one tries to incorporate a microscopic model of the sources in a Lorentz-invariant manner, and so I kept arguing that, at bottom, his 'ideal' was not superior to quantum mechanics. Bell baffled me by rejecting these (valid) objections as irrelevant. It is only

* Presented at the Symposium on Quantum Physics in memory of John Stewart Bell, CERN, May 2–3, 1991.

recently that I came to recognize the point I now believe he sought to make. I would now put it this way.

Imagine someone handing Ampère a sheet of paper bearing Maxwell's equations and the Lorentz force law, and telling Ampère that these describe all the phenomena that he is investigating in his laboratory, and much beyond that, and further, that the crucial new concept in these equations are the symbols $\mathbf{E}(\mathbf{r}, t)$ and $\mathbf{B}(\mathbf{r}, t)$. Armed only with this cryptic message, and his knowledge of Newtonian dynamics, Ampère could then have deciphered precisely what is meant by the electromagnetic field. The essential point is that the new theory, Maxwell–Lorentz electrodynamics, defined its revolutionary basic concepts, *by its equations and without any scholastic crutches*, in terms of more primitive *pre-existing* concepts.

Quantum mechanics does not do this. If one imagines handing the Schrödinger equation to Maxwell, and telling him that it describes the structure of matter in terms of various point particles whose masses and charges are to be seen in the equation, this knowledge would not, by itself, enable Maxwell to figure out what is meant by the wave function. Eventually he would need help: 'Oh, I forgot to tell you that according to Rabbi Born, a great thinker in the yeshiva that will flourish in Göttingen in the early part of the 20th century, $|\Psi(\mathbf{r}_1 \ldots \mathbf{r}_N, t)|^2$ is the joint probability for finding these charges at the indicated positions at time t, and its evolution is determined by the differential equation I revealed to you. But before you rush to the conclusion that Rabbi Born's interpretation answers all your questions, you should reflect on a very profound observation by his great contemporary, Rabbi Dirac of Cambridge, to wit, that if $|\Psi_1(t_0)|^2, \ldots, |\Psi_K(t_0)|^2$ are probabilities of physical interest, then any linear superposition $\Xi(t_0) = \sum c_n \Psi_n$ is, evidently, a valid initial condition for the differential equation, and hence $|\Xi(t)|^2$ must again be a physically realizable probability distribution'. Quite a Talmudic mouthfull.

This contrast between the manner in which quantum mechanics is formulated, when compared to its great revolutionary predecessors, was, I now believe, an essential motive for Bell's conviction that the theory is fatally flawed. And this contrast stimulated his last publication, *Against 'measurement'*, which will be my principal focus [3]. A large part of this article is a biting critique of the treatment of the measurement problem in my 1966 textbook (and in a recent elaboration of it [4]). That he saw fit to devote so much space to decrying my efforts is the highest compliment I have ever been paid.

Against 'measurement' displays all the subtlety and verve that characterized Bell's spoken and written word. Recall the opening passage:

> Surely, after 62 years, we should have an exact formulation of some serious part of quantum mechanics. By 'exact' I do not of course mean 'exactly true'. I only mean that the theory should be fully formulated in mathematical terms, with[1] *nothing left to the discretion of the theoretical physicist*... By 'serious' I mean that some substantial fragment of physics should be covered. Non-relativistic 'particle' quantum mechanics, perhaps with the inclusion of the electromagnetic field... I mean too by 'serious', that 'apparatus' should not be separated off from the rest of the world into black boxes, as if it were not made of atoms and not ruled by quantum mechanics.

This is not a naive plea for a theory that would provide a perfect depiction of some 'serious fragment' of the physical world, let alone a 'theory of everything', the holy grail that so many physicists have chased in recent years, but a call for a theory *formulated in a logically precise manner.*[2] And, true to form, in the same breath as he set down his own preconditions for a satisfactory theory he excoriated the 'the most sure-footed of quantum physicists, those who have it *in their bones* ... [and] become a little impatient with nitpicking distinctions between theorems and assumptions ... and are likely to insist that ordinary quantum mechanics is just fine "for all practical purposes". I agree with them about that: *ORDINARY QUANTUM MECHANICS (as far as I know) IS JUST FINE FOR ALL PRACTICAL PURPOSES.*' For this last phrase Bell coined a marvelously sardonic acronym:

FOR ALL PRACTICAL PURPOSES = FAPP,

which he embellished into a new species of 'theorem':

PROVEN TO BE CORRECT FAPP = QED FAPP.

FAPP and QED FAPP seem destined to become permanent additions to the lexicon of theoretical physics.

Bell held that if quantum theory is to be precise, in the sense just described, then not only must FAPPish arguments be eradicated from its formulation, but the language in which the theory is formulated must be cleansed:

[1] Emphasis added.
[2] Bell also emphasized that 'it is not mathematical but physical precision with which I [am] concerned ... I am not squeamish about delta functions. From the present point of view, the approach of von Neumann's book is not preferable to that of Dirac's.'

> ...however legitimate and necessary in application, [the following words] have no place in a *formulation* with any pretension to physical precision: *system, apparatus, environment, microscopic, macroscopic, reversible, irreversible, observable, information, measurement*... on this list of bad words ... the worst of all is 'measurement'.

And from that point on he always pinned stigmata – quotation marks – onto *measurement.* As he put it to me in Amherst, when we teach people Maxwell's theory, we don't drown the presentation of the theory *per se* in lengthly discussions of arcane minutiae about the use of antennas to measure properties of the ionosphere, etc.; we let the theory speak for itself.

On further reflection, I now think that in arguing his case Bell exaggerated, oversimplified the historical development of both classical electrodynamics and quantum mechanics, and as a result arrived at a definition of 'understanding' that may be too rigid in that it reflects a presumption – a prejudice, if you will – as to what constitutes physical reality. Maxwell and his contemporaries did not present the theory only in terms of the equations so familiar to us, but also employed intricate and highly sophisticated models of the vacuum – the aether – that were crucial to their mode of thought.[3] Long ago, when I was an undergraduate at McGill University, Professor Norman Shaw lectured from notes he had taken while listening to a J.J. Thomson lecture from notes Thomson had taken as a student of Maxwell's! I can assure you that one cannot teach Maxwell's theory in this 'authentic' way to 20th century students because they do not have the background needed to cope with wave

[3] See ref. [5]. Professor Peierls brought my attention to the fact that in the great 1864 paper 'A Dynamical Theory of the Electromagnetic Field' Maxwell emphasized that these mechanical and hydrodynamical models were 'to be understood as illustrative, not explanatory', which implies that by this time Maxwell was no longer wedded to such models. This has led me to peruse *The Scientific Papers of James Clerk Maxwell* (Cambridge University Press, 1890). The above quote is to be found on p. 564, Vol. I. As Peierls acknowledged, Maxwell held contradictory views regarding the aether, which is hardly astonishing. At an earlier point in the same paper (p. 527) he 'assumes that in that space [i.e., in the neighborhood of electric and magnetic bodies] there is matter in motion, by which the observed electromagnetic phenomena are produced'. One of Maxwell's last papers, written between 1877 and his death in 1879, is a semi-popular discussion of the aether, in which he emphasizes that no mechanical model could account for dissipationless propagation, but he nevertheless asserts that 'whatever difficulties we may have in forming a consistent idea of the construction of the aether, there can be no doubt that the interplanetary and interstellar spaces are not empty, but occupied by a material substance or body, which is certainly the largest, and probably the most uniform, of which we have any knowledge' (p. 775, Vol. II). A certain resemblance to speculations regarding dark weakly interacting matter can be noted here!

propagation in conducting fluids permeated by vortices and other exotic objects. More importantly, a modern (post-Einstein) presentation, in which the vacuum is *really empty*, but is, nevertheless, purported to support the transport of energy and momentum, would anger Maxwell and his contemporaries – they would say that we have embraced an empty mathematical formalism that provides no *understanding* of what is '*actually*' going on. Their indictment of Jackson's *Classical Electrodynamics* would sound quite similar to Bell's against such vassals of the House of Bohr as Landau and Lifshitz.

Furthermore, it is not so that the Schrödinger equation (or its matrix mechanics equivalent) must be supplemented by at least a passable interpretation of the wave function (or the matrix elements) before the theory is of any use. Heisenberg, Pauli, Schrödinger, Dirac and their colleagues had firmly established the power of the theory by doing a host of intricate and highly successful calculations before the Born interpretation and the Uncertainty Principle had been put forward. This is not just a debating point. Under an enormously broad set of situations the physical world marches to a quantum mechanical drummer, and was doing so for eons before any conscious being had worried about whether some wave packet had just collapsed. The spectra of distant stars are identical to those of atoms that we humans torture through terrestrial Stern–Gerlachesque mazes; the pace at which thermonuclear reactions proceed within stellar interiors are set by barrier penetration probabilities that bear a clear relation to those measured in our laboratories, and should a star collapse into a compact object its constituents would do so in faithful obedience to the Pauli Principle even though not even one theoretical physicist was watching. Indeed, what are perhaps the two most important triumphs of quantum mechanics – the demonstration that a finite amount of energy is required to excite any bound system, and that every reasonably well isolated atom of some particular species in a particular quantum state has precisely the same properties wherever it happens to be in the universe – do not hinge on a precise understanding of Ψ. Schrödinger, with his totally wrong conception of what his wavefunction meant, knew that his equation explained these most basic properties of matter. Bell was, therefore, resorting to rhetorical overkill when he wrote that

> It would seem that the theory is exclusively concerned about 'results of measurement', and has nothing to say about anything else. What exactly qualifies some physical system to play the role of 'measurer'? Was the

> wave function of the world waiting to jump for [billions] of years until a single-celled living creature appeared? Or did it have to wait a little longer, for some better qualified system ... with a Ph.D.? If the theory is to apply to anything but highly idealized laboratory operations, are we not obliged to admit that more or less 'measurement-like' processes are going on all the time, more or less everywhere? Do we have jumping all the time?

By taking issue with Bell on this score, I am not saying that I am satisfied with *my* understanding of quantum mechanics. I share Feynman's belief that no one really understands quantum mechanics. But we should not exaggerate our embarrassment; it is ample as it stands. In any event, I agree with Bell that the foundations continue to merit the closest scrutiny.

II

Let me, then, examine Bell's critique of quantum mechanical orthodoxy. I do so as a fairly typical defender of established theology, though as with all venerable creeds what Bell and other heretics call 'defenders of orthodoxy' is not one species of dinosaur. My 'orthodoxy' is not identical to that of Bohr, nor to that of Peierls, to mention two especially eminent examples [6]. Hence I must state *my* definition of 'orthodoxy'.

I. The interpretation of the scalar product as a probability amplitude cannot be derived from the Schrödinger equation. Various strategies have been advocated for coping with this conundrum. Those that modify the Schrödinger equation and/or the content of the wave function I call 'heretical'. My strategy is to seek a demonstration that the statistical interpretation is *compatible* with the unitary evolution of state vectors under well-specified circumstances.[4] What could be called 'canonical orthodoxy' is to add von Neumann's wave packet reduction postulate to the axiomatic basis, and it is my principal objective to avoid this. Indeed, a central theme in the debate is whether the reduction

[4] In 'Against' (p. 36), Bell asserts that 'reading [my book] reveals other ambitions'. That is not so. The opening page (p. 165) of the chapter on measurement has the following statement (in italics): 'Here, as nowhere else, it is crucial to keep one's eye firmly focused on the ultimate objective: *To verify the consistency of the mathematical formalism with the interpretation of the scalar product as a probability amplitude.*'

postulate is, or is not, indispensable to a demonstration of compatibility between the interpretation and the superposition principle.

II. A *pure* quantum state, represented by a ray in the Hilbert space, describes an ensemble of identically prepared systems.

This also requires some elaboration. First, I see no need to apologize for this definition. Physics, in contrast to difficult pursuits, is the study of *reproducible* phenomena. In the microscopic realm it is an *empirical fact*, learned without *any* help from theory, that only the behavior of ensembles is, in general, reproducible, whereas that of individual systems is not. At one time it was possible to entertain the conjecture that there are hidden variables which, when discovered, will remove the need for statistics at a fundamental level, but the experiments inspired by the Bell inequality have shut that escape hatch as they rule out all but non-local hidden variable theories. Hence a statistical theory of the microcosmos is all that theoretical physics *should* seek. Second, some critics of 'orthodoxy' argue that by assuming that the state vector does not describe an individual system, but only ensembles, one has, by definition, done away with the measurement problem. I do not agree. Important problems remain even if one accepts the irreducibly statistical character of quantum mechanics. Furthermore, the treatment of these remaining problems that I shall sketch is not entirely satisfactory. In short, as I see it, statement II does not offer peace with honor between heresy and 'orthodoxy'.

III. I agree with Bell that the '... "apparatus" should not be [treated] as if it were not made of atoms and ruled by quantum mechanics.' Here my 'orthodoxy' differs from Bohr's, but not from that of Peierls.

This definition of 'orthodoxy' then requires the confirmation of the following statements:

A. Whenever an ensemble $\{S\}$ of systems S is prepared in the same manner, upon being sent through a measurement apparatus $\mathcal{A}$ one finds a joint probability distribution that correlates each state of

$\{\mathcal{S}\}$ measured by $\mathcal{A}$ with one macroscopically distinguishable state of $\mathcal{A}$, so that a macroscopic observation of the latter suffices to determine the former.

B. If a *single* system $\mathcal{S}$ is sent through $\mathcal{A}$, one cannot predict which of the several outcomes in the probability distribution will actually occur, but in each and every trial *one* definite outcome will be observed.[5]

For the sake of definiteness, let $\mathcal{S}$ be a neutral atom and $\mathcal{A}$ a Stern–Gerlach set-up furnished with detectors that register whether or not an atom is in the neighborhood of the classical trajectory $\mathcal{T}_m$ as determined by the z-projection of the atom's magnetic moment, $\mu_B gm$, where μ_B is the Bohr magneton.[6] After entry into $\mathcal{A}$ the atom's wave function is $\eta_m \phi_m(\mathbf{R})$, where η_m is the internal state and $\phi_m(\mathbf{R})$ the center-of-mass wave function which goes to zero rapidly as $\mathbf{R}$ recedes from $\mathcal{T}_m$. Each detector $\mathcal{D}_m$ is a system – such as a surface on which the atoms are deposited for subsequent observation as in the original experiment, or a counter that changes its state when traversed by a magnetic moment. $\mathcal{D}_m$ is to be incorporated into the Schrödinger equation; it has variables designated by x_m, and a ground state $\chi_0(x_m)$. The essential property of the detectors is that they only have a short-range interaction with the atoms, so that when atoms in the state η_m pass through $\mathcal{A}$ they can *only* influence $\mathcal{D}_m$.

When $\eta_m \phi_{\text{inc}}(\mathbf{R})$ is incident, the state that emerges is

$$\Psi_m = \eta_m \phi_m(\mathbf{R}) \chi_E(x_m) \prod_{n \neq m} \chi_0(x_n), \tag{1}$$

where χ_E describes the excited state of $\mathcal{D}_m$. For a detector of high efficiency, the overlap between χ_0 and χ_E is small, and for brevity's sake we shall treat it as if it vanished. If the initial state of the atoms is the arbitrary *pure* state $\sum c_m \eta_m$, the superposition principle tells us that the combined state of $\mathcal{S}$ *and* $\mathcal{A}$ that emerges is

$$\Psi = \sum_m c_m \Psi_m. \tag{2}$$

[5] Assuming perfect counting efficiency, of course.

[6] The Stern–Gerlach paradigm has become antiquated; it should be replaced by examples drawn from quantum optics. See, for example, refs. [7] and [8].

The central objective of the exercise is then to demonstrate that the $|c_m|^2$ are probabilities for finding atoms with the eigenvalues m of J_z.

If $\mathcal{A}$ is properly designed, it is an amplifier whose output is a superposition of states of the detecting systems, each of which has no overlap in configuration space (in this example, the space of the x_m) with any of the others. To be more precise, if $\mathcal{O}$ is any observable related to the atomic center-of-mass and/or the detectors, and if $m \neq m'$, the matrix element $\langle \mathbf{R}, x_m | \mathcal{O} | \mathbf{R}', x'_{m'} \rangle$ is assumed to tend rapidly to zero when the separation between the coordinates in the bra and ket grow to macroscopic proportions. We call such observables *macroscopically local.* For any such $\mathcal{O}$ this apparatus has established a one-to-one correlation between the states of the micro-system $\mathcal{S}$ and macroscopically distinguishable states of the apparatus. If $\mathcal{A}$ cannot produce such an outcome it is a contraption, however useful or intriguing, that does not accomplish the measurement task, but such a failure would put no onus on quantum mechanics.

Obviously (2) is a pure state. But if all the observables $\mathcal{O}$ at our disposal are macroscopically local, this pure state Ψ cannot be distinguished from the mixture obtained by deleting all interference terms between states of differing m from the density matrix:

$$\hat{\rho} = \sum_m |c_m|^2 \phi_m(\mathbf{R}) \phi_m^*(\mathbf{R}') \chi_{\mathrm{E}}(x_m) \chi_{\mathrm{E}}^*(x'_m) \prod_{n \neq m} \chi_0(x_n) \chi_0^*(x_n). \qquad (3)$$

This 'butchered' density matrix has the structure of a *classical* joint probability distribution, without mysterious interference terms, and displays the one-to-one system–apparatus correlations demanded by requirement A. Hence it identifies the $|c_m|^2$ as the probabilities that $\mathcal{S}$ will be found to 'be in' the mth state η_m with the indicated eigenvalue. Furthermore, in (3) each term has one, and only one, detector excited, so that a sufficiently dilute beam will only cause a single detector to trigger at a time.[7] In short, both the requirements A and B have been met. Indeed, as Bell once put it [9]:

> In quantum measurement theory such elimination of coherence is the philosopher's stone. For with an incoherent mixture specialization to one of its components can be regarded as a purely mental act, the innocent selection of a particular subensemble, from some total statistical ensemble ...

[7] Note that this conclusion does not follow from the most stripped-down description of the Stern–Gerlach experiment, in which $\mathcal{A}$ is represented only by the center-of-mass $\mathbf{R}$.

I must emphasize that while I agree with this statement,[8] it turns out that Bell was in his sarcastic mode when he wrote it in 1976, and did not mean it to be taken at face value.[9] In *Against 'measurement'* Bell called this the LL (for Landau and Lifshitz) jump – '... [the] jump ... of a "classical" apparatus into an eigenstate of one of its "readings" ... the *spontaneous* jump of a macroscopic system into a definite macroscopic configuration.' Here the emphasis on 'spontaneous' and the repetitive use of 'jump' is a mystery to me: when the roulette wheel stops in Monte Carlo, and assuming no chicanery, is that not also a spontaneous jump into a definite state from what was a probability distribution before the wheel was spun? Why is the inference I have drawn from $\hat{\rho}$ different from what one would infer from ruinous nights spent in casinos? Admittedly, there is a residual coherence in the quantum case that has no Monte Carlo counterpart, but this is not observable FAPP, though the validity of this last statement does not hold forever, strictly speaking, a point to which I now turn.

The time T_C over which the imposter $\hat{\rho}$ can masquerade successfully as the true density matrix with its interference terms is very sensitive to the extent to which a realistic description of the apparatus $\mathcal{A}$ is included in the Schrödinger equation. At the most naive extreme, when only the atom's center of mass $\mathbf{R}$ is considered to be $\mathcal{A}$, with the differing trajectories $\mathcal{T}_m$ playing the role of pointer, it is relatively straightforward to add fields that combine the separated beams and to re-establish coherence – and by that token, to simultaneously destroy the ability of $\mathcal{A}$ to effect any measurement. In this case T_C is short.

But T_C expands enormously as soon as a more realistic description of $\mathcal{A}$ is included in the Schrödinger equation. Consider, for example, the inclusion of the surface on which the atoms are deposited after traversal of the magnet, and assume, for simplicity's sake, that it is legitimate to treat this surface as being in its ground state before impact (see [10]). The initial state of surface

[8] Essentially the same statement appears in my text (p. 189): '... the conventional statistical interpretation ... follows [from $\hat{\rho}$] by employing concepts familiar to us on the macroscopic (or classical) level of perception. To be sure, a reduction does occur in the statistical distributions that arise from $\hat{\rho}$, but there is nothing novel to quantum mechanics to this. In classical probability theory the state of a coin following the toss is, say, "heads", whereas before the toss it was 50% "tails" and 50% "heads"'.

[9] I learned this when I cited his statement in publicly presenting my position at Amherst, at which point Bell impatiently interrupted and said that I had failed to grasp his point! That was indeed so, and remains so, despite Abner Shimony's passionate and patient efforts to instruct me.

plus atom is in the continuum, and a transition must then occur into a state in which the atom is localized about a point X_m close to the intersection of the classical trajectory $\mathcal{T}_m$ with the surface. In general, the transition will be accompanied by non-localized excitations of the surface (e.g., phonons), and result in capture into a very narrow level out of which the atom can only tunnel to a neighboring position after a very long time. That $\hat{\rho}$ is an imposter can only become known when the state just described attains detectable coherence with one in which the atom was deposited originally at a different location $X_{m'}$, and which is macroscopically separated from X_m. But quantum mechanical coherence requires overlap in *configuration* space, and not just in 3-space. The wave functions for the two situations, localization about $X_{m'}$ and X_m, involve a very large number of coordinates describing the local environments of the surface as well as excitations that may have propagated afar. Even if all but a small portion of these variables eventually come to overlap appreciably (e.g., through diffusion of the localized atoms by successive tunnelings) there will still be negligible coherence if the other variables (describing, in this example, non-localized excitations) fail to overlap. As we see, time scales of enormous extent characterize the interval over which coherence is undetectable, and it seems impossible to devise some external intervention (akin to the recombining fields in the primitive Stern–Gerlach illustration) that could enforce coherence.

To summarize, I claim that

- the statistical distribution without interference terms between distinct outcomes, as required for a meaningful interpretation, is indistinguishable from the density matrix that evolves in accordance with the *unitary* Schrödinger evolution once a reasonably realistic description of the measurement process is incorporated into the dynamics, for all observables pertaining to $\mathcal{S}$ and $\mathcal{A}$ that are macroscopically local;
- these interference terms can only reappear, if at all, in some exceedingly entangled fashion if we wait for a time that is enormous compared to macroscopic (and not atomic) time scales.
- QED FAPP.

The time T_C over which coherence is undetectable therefore plays a role similar to that of the Poincaré recurrence time in classical statistical mechanics.

Admittedly, it is somewhat unsettling that the argument has a flavor reminiscent of that which concludes that classical statistical mechanics can be used with confidence for FAPPish endeavors, for we know that statistical mechanics is only a superstructure on a fundamental theory, whereas quantum mechanics is supposed to be fundamental.

The restriction to macroscopically local apparatus observables also requires a comment. In contrast to Dirac's usage, I restrict the term 'observable' to entities that actually occur in nature, or for that matter, that one might imagine to occur in nature. In Dirac's terminology, all Hermitian operators are observables. Apart from a set of measure zero (which includes those of physical interest), Dirac's 'observables' are macroscopically *non*-local. The theory should not be expected to have an interpretation in terms of our own (i.e., macroscopic) concepts unless the Hamiltonians admitted to the Schrödinger equation are constrained to reflect the real world from which we have gleaned our interpretive conceptions – there is no reason why the theory should produce outcomes of which we can make sense if unphysical interactions between $\mathcal{S}$ and $\mathcal{A}$ are put into Schrödinger's equation of motion.[10]

III

Let me now try to render Bell's criticisms of the above argument. First, it was not focused primarily on the replacement of the true density matrix by the butchered imposter $\hat{\rho}$: 'I am fully convinced of the practical elusiveness, even the absence FAPP, of interference between macroscopically different states.' What did vex him was (1) that I did not seem to understand the difference between *and* and *or*; (2) that my argument showed 'some conceptual drift' which had, perhaps unwittingly, led to tacit use of the reduction postulate even though my proclaimed purpose was to shun it; and (3), that like all those unwilling to abandon 'orthodoxy' (of whatever strain), I resorted to what Bell viewed as an unacceptable 'shifty split' of the world into a part that quantum mechanics purports to describe and a remainder that falls outside the theory's realm.

[10] As already pointed out in my book, superfluids possess macroscopically nonlocal observables, and could (indeed have now) been used to display interference over macroscopic separations. But if such a device were added to $\mathcal{A}$, it would not invalidate the interpretation of quantum mechanics, but only render the contraption in question useless for the measurement that was the original goal.

First, then, the and/or issue. Bell argued that if the various terms in (3) correspond to *distinct alternatives*, what should be demonstrated (at least at the QED FAPP level) is not that the outcome is the *sum* of these terms, as in $\hat{\rho}$, but rather:

$$\begin{aligned}\Psi\Psi^* &\rightarrow |c_1|^2\eta_1\eta_1^*\phi_1\phi_1^*\cdots \\ &\rightarrow \text{or } |c_2|^2\eta_2\eta_2^*\phi_2\phi_2^*\cdots \\ &\rightarrow \text{or } \cdots.\end{aligned}$$

This might seem to be a devastating blow. But in $\hat{\rho}$ the different terms have no overlap in the space of the apparatus variables (i.e., **R** and the x_m); moreover, if a conventional probability distribution (e.g., of a population's incomes) breaks up into two or more non-overlapping pieces, these can always be added together, normalized to unity, and the sum plotted as a function of the variables that label the individual members of the ensemble. After this is done, it is still legitimate to say that this sum of terms represents the probability that a member will fall into one *or* another of the subensembles. As this is obvious, Bell's criticism could not be so superficial. To my understanding, his point was that if *only* $\hat{\rho}$ is provided, that does not imply that $\hat{\rho}$ represents the probability distribution for a set of *distinct* events – as in the case of a beam sufficiently dilute so that only one atom at a time passes through $\mathcal{A}$. Put another way, the function $\hat{\rho}$ could equally well represent an indivisible continuum, as in Schrödinger's original misconception that $|\Psi|^2$ is a continuous charge distribution. Indeed, I seem to recall that Bell made this point to me in a 1964 correspondence while I was writing my book, and that this stimulated me to end my chapter on measurement with the disclaimer 'we have not attempted to reconcile the fact that the theory only makes statistical predictions with our observation of individual, solitary, events.'

Recently I have realized that this concedes too much: that an elementary exercise shows that (see Appendix) when nothing but the Schrödinger equation is used to describe a sufficiently dilute beam interacting with, say, a Stern–Gerlach setup, the detecting system will record isolated hits – that the number of atoms trapped on the detecting surface climbs in very brief and well separated intervals by unit steps. That is to say, by using second quantization, which provides a more realistic description of the state of the system $\mathcal{S}$ before measurement (as compared to a more realistic description of $\mathcal{A}$, which heretofore has been my defensive tactic), one can show that the density matrices under discussion are probability distributions for isolated events. What I do not know is whether this argument suffices to change any minds, or whether it misses a point that is too subtle for me to have grasped thus far.

Next comes the second point – whether the reduction postulate slipped in through the back door, or as Bell put it: 'The mystery is then: what has the author [KG] actually derived rather than assumed?' As my objective is only to establish consistency between the interpretation and the unitary dynamics, my argument is, by definition, circular. The question is what is on the circle, and what is not, and how close does the circle come to being a tautology.

The argument leading to $\hat{\rho}$ does not rely on an interpretation of the atoms' internal wave functions η_m beyond what can be inferred directly from the Schrödinger equation, namely, that these functions contain all the information that the theory can provide regarding the internal structure and dynamics of the atoms. Concerning the center-of-mass functions $\phi(\mathbf{R})$, by solving the Schrödinger equation one can demonstrate that ϕ vanishes unless $\mathbf{R}$ is in close proximity to the classical trajectory $\mathcal{T}_m$ appropriate to the classical magnetic moment $\mu_B gm$. *Therefore the only assumption concerning the interpretation of wave functions that is needed is that the atom is never in places where* $\phi(\mathbf{R})$ *vanishes*, for that suffices to tell us where, *on a macroscopic length scale*, atoms in the subensembles labeled by m strike the detector surface, and from that the (eigen)values of the magnetic moment and the corresponding probabilities $|c_m|^2$. The statement in italics is thus the suspect on the circle; but it is a much weaker assumption than the full-fledged Born edict, which governs all length scales dealt with by nonrelativistic quantum mechanics. To some extent, Bell acknowledges that my argument does accomplish something: 'This is important, for it shows how, FAPP, we can get away with attributing definite classical properties to "apparatus" while believing it to be governed by quantum mechanics. But a jump [i.e., reduction] assumption remains . . . KG derives, FAPP, the LL jump from assumptions about the shifted split [between $\mathcal{A}$ and the rest of the world], which includes the [von Neumann] jump there.' Unfortunately, I do not understand the last phrase – where I unwittingly used the von Neumann jump, and must leave the matter there.

Finally, then, there is the shifty split, which Bell puts as follows in his indictment: 'The basic structure of KG's world is then $W = \mathcal{S} + R \ldots$ where $\mathcal{S}$ is the quantum system, and R is the rest of the world – from which measurements on

$\mathcal{S}$ are made. When our *only* interpretive axioms are about measurement results we absolutely *need* such a base R... there can be no question then of identifying the quantum system $\mathcal{S}$ with the whole world W. There can be no question – without changing the axioms – of getting rid of the shifty split ... But there is something which can and must be done – to analyze theoretically not *removing* the split, which cannot be done with the usual axioms, but of *shifting* it. This is taken up in KG's Chapter 4 ... there is then some conceptual drift in the argument ... it is supposed that the LL jump really takes place. The drift is away from the "measurement" (... external intervention ...) orientation of orthodox quantum mechanics towards the idea that systems, such as $\mathcal{S} + \mathcal{A}$ [with $\mathcal{A}$ sufficiently realistic], have intrinsic properties – independent of and before observation. In particular, the readings of experimental apparatus are supposed to be really there before they are read.... The resulting theory would be one in which some "macroscopic" "physical attributes" *have* values at all times, with a dynamics that is related somehow to the butchering of ρ into $\hat{\rho}$... But the retention [by KG] of the vague word "macroscopic" reveals limited ambition as regards precision ... To avoid the vague "microscopic"–"macroscopic" distinction – again a shifty split – I think one would be led to introduce variables which *have* values [at all times] even on the smallest scale.'

I plead guilty to some of these charges:

- I think it has been, and continues to be, a blessing that we need not take the whole world W into account if we are to understand the microcosm. Nothing I have been able to understand has swayed me to accept Bell's assertion that 'a serious formulation [of quantum mechanics] will not exclude the big world outside the laboratory'.
- Nor has anything I have been able to understand led me to lose sleep over the 'shifty split', or over demonstrations that only attain the QED FAPP level of rigor – to paraphrase Bell, I am not squeamish about QED FAPP. The theory itself tells us under what circumstances the classical approximation is, FAPP, valid for some particular system. Once the description of the apparatus crosses a reasonably well-defined threshold of authenticity – a threshold that is not determined by the whim of some theoretical physicist but by the theory itself, the

statistical interpretation becomes compatible with the mathematical formalism; further embellishments do not, FAPP, alter the situation: the 'shifty split' becomes, so to say, 'translation' invariant FAPP.

- I believe that the argument I have given shows that once the above-mentioned threshold is crossed the system $\mathcal{A}+\mathcal{S}$ does have properties 'independent of and before observation' by still more complex systems, including, for example, systems that draw a CERN Grade 14 salary.[11]
- Given the enormous range of distances and times that separate everyday life from the microscopic world, I find it most impressive that quantum mechanics can tell us when, FAPP, the 'shifty split' is permissible, and thus, when our gross, human-scale intuition will, by and large, not lead us astray.

On the other hand, I do, and with shame, plead guilty to having presented an argument that is slipshod – that there is indeed a conceptual drift which reflects an inadequate understanding of what I 'have actually derived rather than assumed'. That is, I agree with Bell that I have a too 'limited ambition as regards precision', but I do not agree that the 'smoking gun' is a 'vague' distinction between macroscopic and microscopic. The smoking gun is that I do not understand properly how much of the mathematical formalism – operators, expectation values, etc. – it is legitimate to use in a demonstration of consistency between the theory's mathematical formalism and its interpretation.

IV

Finally, I return to the question whether quantum mechanics 'carries within itself the seeds of its own destruction'. If this verdict is correct, it would mean that theoretical physics is in deep crisis, a condition that usually leads to a profound breakthrough. But we must then ask what, precisely, is the crisis we are faced with?

[11] In this connection, see the argument on p. 183 of my text which concludes that it does not matter whether we 'look' at the detectors *provided* there is an interaction between them and the atoms.

In addressing this question, it is useful to recall two very different crises that beset physics at the beginning of this century. The first was Einstein's realization that the equivalence principle implied that one could not construct a theory of gravitation compatible with the principle of relativity by restricting the space–time transformations to those that form the Lorentz group. This, then, was a deep conceptual flaw in the fabric of existing theory formulated in a mathematically well-defined fashion, and the evident need to remove this flaw led directly (at least in hindsight) to general relativity. The second example was the inability of classical physics to reconcile the compelling hypothesis that matter is composed of atoms with crucial *qualitative* properties of matter: that all atoms and molecules of any particular species are identical, and do not alter their properties under a wide range of physical and chemical processes; and that atoms and molecules behave as if they had a fixed and quite small number of degrees of freedom [11]. Classical concepts imply that any species of molecules should be expected to display a continuum of structures and forms, and that a fixed number of molecular degrees of freedom is incompatible with any internal structure whatsoever for atoms, which is contradicted by the qualitative features of atomic spectra. As we now know, quantum theory removes both of these conundrums, but at the time of the theory's invention the crisis was only understood in another guise – that classical theory could not handle the interaction between matter and the radiation field, a dilemma that had been formulated in a precise mathematical form by Wien, Rayleigh and Jeans. That classical theory had catastrophic implications for the constitution of atoms was barely appreciated by Planck and others at the turn of the century, and played no crucial role in the development of quantum theory until Bohr's work 13 years later.

The lesson that could be drawn from this, and from the development of general relativity, is that a crisis will only fertilize new insights if it is formulated in a mathematically incisive manner. This conclusion has an echo in Bell's insistence on having quantum mechanics '*fully* formulated in mathematical terms, with nothing left to the discretion of the theoretical physicist', but what it really calls for is a critique of the 'orthodox' theory fully formulated in mathematical terms with nothing left to the discretion of the critic.

That such a critique is not yet in hand was acknowledged by Bell in the same breath as he held out the hope that the failings he saw in the 'orthodox' theory held out great prospects for the future of physics:

> Suppose for example that quantum mechanics was found to *resist* precise formulation. Suppose that when formulation beyond FAPP is attempted, we find an unmovable finger pointing outside the subject, to the mind of the observer, to the Hindu scriptures, to God, or even only to Gravitation? Would not that be very, very interesting?

But the vision of the future that he conveyed was a mixed signal. For example, in the 1960s he [12]

> look[ed] forward to a new theory which can refer meaningfully to events in a given system without requiring 'observation' by another system. The critical test cases requiring this conclusion are systems containing consciousness and the universe as a whole ... [we] share with most physicists a degree of embarrassment at consciousness being dragged into physics, and share the usual feeling that to consider the universe as a whole is at least immodest, if not blasphemous ... It seems likely ... that physics will again adopt a more objective description of nature long before it begins to understand consciousness, and the universe as a whole may well play no central role in this development.

On the other hand, he also thought that there are just two options for completing the theory [13]:

> Either the wave function, as given by the Schrödinger equation, is not everything, or it is not right ... [thus] two roads are open towards a precise theory ... Both eliminate the shifty split. The de Broglie–Bohm ['pilot wave'] theories retain, exactly, the linear wave equation, and so necessarily add complementary variables to express the non-waviness of the world on the macroscopic scale. [In] the Ghirardi–Rimini–Weber type theories[12]... the linear Schrödinger equation has to be modified ... by a mathematically prescribed spontaneous collapse mechanism.

But neither of these theories include 'the big world outside the laboratory', nor give any discernible hint of why such an inclusion might be necessary, unless the 'spontaneous collapse mechanism' is, in some not yet understood sense, actually brought about by the existence of 'the big world'.

I, for one, remain unpersuaded that a crisis of major proportions has been identified, but at the same time prudence compels me to be open to the possibility that such a crisis has been glimpsed, though perhaps only subliminally, by Bell and his great predecessors: Einstein, Schrödinger, de Broglie, Bohm and Wigner. This is a roster which should give anyone pause. To my understanding,

[12] See the contribution by G.C. Ghirardi to this volume.

however, *this crisis has not yet been formulated with sufficient precision to facilitate the birth of a great offspring.* Furthermore, with Bell I would wager that [14] 'what seems ... likely is that new ways of seeing things will involve an imaginative leap that will astonish us'. And as I told John in our last conversation, I would also wager that the new theory will be even further removed from our philosophical and conceptual prejudices than is quantum mechanics, and that he would then pine for the time when he had 'good old' quantum mechanics to kick around.

Appendix

The language of second quantization is best suited to analyzing whether the 'orthodox' theory can account for the observation of isolated events. Let $\psi_m(\mathbf{R})$ create an atom in the state η_m with its center of mass at $\mathbf{R}$, and

$$\psi(\mathbf{R}) = \sum_m c_m \psi_m(\mathbf{R}).$$

Next, introduce a set of wave packets $u_n(\mathbf{R}, t)$ which move with the mean velocity of the incident beam, take the time T_{tr} to traverse the apparatus $\mathcal{A}$, and perform the traversal in the interval $n\tau < t < (n+1)\tau$, where $\tau \gg T_{\mathrm{tr}}$. Finally, define the one-atom creation operators

$$\Phi_n^\dagger(t) = \int \mathrm{d}^3 R u_n(\mathbf{R}, t) \psi^\dagger(\mathbf{R}). \tag{4}$$

We can now construct states of the system plus apparatus that describes very dilute atomic beams. For $t < 0$, such a state is

$$|\mathcal{S} + \mathcal{A}\rangle = \prod_{n=0}^{\infty} \Phi_n^\dagger(t)|0\rangle, \tag{5}$$

where $|0\rangle$ is a product of the detector's ground state and the state with no atoms. For $t > 0$ this state will evolve in accordance with the interactions between $\mathcal{S}$ and $\mathcal{A}$. At any time, no more than one atom is on its way through $\mathcal{A}$ towards impact on the detecting surface – the state always has either the eigenvalue 0 or 1 of the operator

$$N_{\mathcal{A}} = \sum_m \int \mathrm{d}^3 R \psi_m^\dagger(\mathbf{R}) \psi_m(\mathbf{R}), \tag{6}$$

where the integration is restricted to the volume of $\mathcal{A}$ *excluding* the surface of the detector.

The interaction between the atoms and the detecting surface is

$$V = \sum_m \int d^3R\, d^3x\, v(\mathbf{R}-\mathbf{x})\psi_m^\dagger(\mathbf{R})\psi_m(\mathbf{R})\Theta(\mathbf{x}), \tag{7}$$

where $\Theta(\mathbf{x})$ is an operator that causes a change of state of the detector emanating from a point $\mathbf{x}$ 'on' its surface, and v is a very short range, highly attractive effective interaction that traps atoms onto the surface. If we define N_{surf} by the same expression as $N_{\mathcal{A}}$, but with the integration running only over the detecting surface, the eigenvalue of N_{surf} increases by 1 in every interval of duration τ. Furthermore, if sufficiently short packets u_n are used, the time over which this eigenvalue changes will be far briefer than T_{tr}.

ACKNOWLEDGMENTS

I am especially indebted to David Mermin, Abner Shimony and Victor Weisskopf for many lengthy and instructive discussion about these issues. I have also benefited greatly from conversations with Sidney Coleman, Roy Glauber, Jeffrey Goldstone, Hermann Haus, Yuri Orlov, the late Rudolf Peierls and the late Donald Yennie. When writing on this topic, however, it is essential to emphasize that the author is fully responsible for all statements in the text that are not unambiguously ascribed to others.

I am most grateful to many old friends at CERN, and in particular Daniele Amati and John Ellis, for having given me the opportunity to contribute to this memorial to John Bell, a celebration of what is finest in the human spirit; to the Theory Secretariat for their habitual but nonetheless inimitable support; and to Rufus Neal and Cambridge University Press for their sustained commitment to the publication of this volume.

A second, quite different, and I believe more credible response to John Bell's critique of my 1966 treatment of the statistical interpretation of quantum mechanics appears in *Epistemological and Experimental Perspectives in Quantum Physics*, 7th Yearbook Institute Vienna Circle, eds. D. Greenberger, W.L. Reiter and A. Zeilinger (Kluwer, Dordrecht, 1999).

REFERENCES

[1] J.S. Bell and M. Nauenberg, The moral aspect of quantum mechanics, in *Preludes in Theoretical Physics*, A. de-Shalit, H. Feshbach and L. Van

Hove (eds.), (North-Holland, 1966), pp. 279–86. With the exception of ref. [3], this and all other articles by Bell on the foundations of quantum mechanics are reprinted in J.S. Bell, *Speakable and Unspeakable in Quantum Mechanics* (Cambridge University Press, 1987). Ref. [3] is included in *Quantum Mechanics, High Energy Physics and Accelerators, Selected Papers of John S. Bell*, M. Bell, K. Gottfried and M. Veltman (eds.), (World Scientific, 1995).

[2] For a sensitive portrait of Bell, see Jeremy Bernstein, *Quantum Profiles* (Princeton University Press, 1991).

[3] J.S. Bell, Against 'measurement', *Physics World*, August 1990, pp. 33–40. All quotations that carry no citation are from this article.

[4] K. Gottfried, *Quantum Mechanics* (W.A. Benjamin, 1966), reissued by Addison-Wesley, 1989; 'Does quantum mechanics describe the "collapse" of the wave function?' presented at *62 Years of Uncertainty*, Erice, August 1989, unpublished. See also M. Cini, 'Quantum theory of measurement without wave packet collapse', *Nuovo Cimento* **73B** (1983) 27. I must point out, as I did in my book, that 'my' treatment is an elaboration of earlier discussions, and in particular, of that in W. Pauli, *Die Allgemeinen Prinzipien der Wellenmechanik* (J. Edwards, Ann Arbor, 1950), pp. 143–154.

[5] E.T. Whittaker, *A History of the Theories of Aether and Electricity, Vol. I, The Classical Theories* (Thomas Nelson & Sons, 1951).

[6] R. Peierls, In defence of "measurement", *Physics World*, January 1991, pp. 19–20.

[7] R.J. Glauber, Amplifiers, Attenuators, and Schrödinger's Cat. *New Techniques in the Quantum Measurement Theory*, Ann. N.Y. Acad. Sc. **480** pp. 336–372.

[8] Y. Yamamoto and H.A. Haus, *Rev. Mod. Phys.* **58** (1986) 1001.

[9] J.S. Bell, On wave packet reduction in the Coleman–Hepp model, *Helv. Phys. Acta.* **48** (1976) 93.

[10] A. Danieri, A. Loinger and G.M. Prosperi, *Nuclear Physics* **33** (1962) 297.

[11] K. Gottfried and V.F. Weisskopf, *Concepts of Particle Physics*, Vol. I (Oxford University Press, 1984), pp. 5–6.

[12] Bell & Nauenberg, ref. [1], p. 284.

[13] J. Bell, Are there quantum jumps? in *Schrödinger, Centenary of a Polymath* (Cambridge University Press, 1987).

[14] Bell & Nauenberg, ref. [1].

8 John Bell and the moral aspect of quantum mechanics

KURT GOTTFRIED AND N. DAVID MERMIN

Many years ago, during a pleasant lunch in the CERN cafeteria, with John's penetrating observations having demonstrated yet again his remarkable grasp of the foundations of physics, Mary Bell quietly said that her husband's secret ambition was to find an apartment on the *Boulevard des Philosophes*, an assertion that John let pass with a smile.

Bell's love of natural philosphy was tempered by the conviction that to earn one's daily bread in both the intellectual and material sense one must have a respectable profession, which for him happened to be ordinary theoretical physics,[1] a *métier* he practised with superior skill on an exceptionally broad front, from the design of acclerators to the symmetries of fundamental interactions. This remarkable mixture of the practical and the contemplative reflected, on the one hand, the pragmatic British tradition in which he was schooled, and on the other, his deepest human needs. John Bell had a consuming commitment to wresting an *understanding* of the natural world from the great theories of physics. He held that a theory that merely succeeded in accounting brilliantly for data, without providing a satisfactory understanding of what it described, should be subject to stringent critical scrutiny, and if such an understanding was found to be unattainable the theory should be expected to crumble, its superficial triumphs notwithstanding. Quantum mechanics was number one on Bell's wanted list, and throughout his life he pursued the culprit and those who were content to rest with the proposition (with which he agreed) that 'ordinary quantum mechanics is just fine for all practical purposes'.

[1] Jeremy Bernstein reports [1] that Bell listed his physics speciality in an official CERN document as 'quantum engineering', (p. 12), and averred that 'I am not like many people I meet at conferences on the foundations of quantum mechanics ... who have not really studied the orthodox theory [and] devote their lives to criticizing it ... I think that means they haven't really appreciated the strengths of the ordinary theory. I have a very healthy respect for it. I am enormously impressed by it.' (p. 85).

This quest for understanding had two distinct aspects. One led Bell to pin down in a precise and powerful way some of the most dramatic features of the revolution in thinking that quantum mechanics forces on us, through penetrating examinations of the possibility of hidden variables theories and the nature of quantum non-locality. The other yielded an unrelenting critique of the conventional formulation and interpretation of quantum mechanics and, by implication, of the very purposes and goals of contemporary theoretical physics.[2]

In two remarkable papers written some 25 years ago, Bell demolished a mythology that had inhibited clear thinking about the meaning of quantum mechanics since the mid-1930s, and reformulated the issues with a clarity and force that focused the energies of many hard-nosed theoretical *and* experimental physicists on questions that lie at the very heart of quantum mechanics.

In the first of these papers [3], Bell examined the old question of whether the states of quantum mechanics can be viewed as ensembles of 'dispersion free' states, specified by additional variables whose values determine precisely the results of *individual* measurements. Since 1932, most physicists had gladly avoided such efforts because of von Neumann's famous theorem asserting that any attempt to embellish the conventional theory with hidden variables must necessarily disagree with some of the theory's quantitative predictions. In retrospect, it would seem that in the intervening decades few could actually have taken the trouble to penetrate von Neumann's 19 page proof since, as Bell pointed out, the hidden variable theories the theorem actually excluded were required to satisfy a superficially plausible but physically unjustifiable – indeed, upon reflection, quite arbitrary – constraint.

Bell then went on to prove a theorem of his own, which reached von Neumann's conclusion without this unacceptable restriction, but which, as Bell pointed out, relied on a much more subtle assumption – 'that so much follows from such apparently innocent assumptions leads us to question their innocence'. His own prohibition of hidden variables, Bell noted, did not apply to theories in which 'the result of an observation may ... depend not only on the state of the system (including the hidden variables) but also on the complete disposition of the apparatus'. This was a 'judo-like' *tour de force*,

[2] Most of Bell's shots across the bow of orthodoxy are collected in ref. [2].

as Abner Shimony subsequently put it, for Bell had found hope for hidden variable theorists in the teachings of Bohr, the high priest of orthodoxy!

Possibly because Bell described it as a 'corollary of Gleason's work', this major result is generally known as the Kochen–Specker theorem (in recognition of a later but independent derivation of essentially the same conclusion). For the sake of clarity this is probably just as well, for 'Bell's Theorem' – the very different content of the second paper [4], has now achieved a level of popular acclaim exceeded by few theorems in the history of mathematics and physics.

What Bell's Theorem demonstrated is that Bohr's lesson is much stranger than even Bohr may have realized. For, in the second paper, Bell proved a no-hidden-variables theorem that applied to theories constrained only by the requirement that how the hidden variables influence the results of observations should not depend on the disposition of the apparatus *far away* from where those observations are performed [5]:

> In a theory in which parameters [hidden variables] are added to quantum mechanics to determine the results of individual measurements, without changing the statistical predictions, there must be a mechanism whereby the setting of one measuring device can influence the reading of another instrument, however remote. Moreover, the signal involved must propagate instantaneously, so that such a theory could not be Lorentz invariant.

This result is often characterized as a proof that no local hidden variables theory can reproduce the results of quantum mechanics. But Bell's Theorem is more than just that. His analysis, which is extraordinarily elementary, has the kind of generality one encounters at the foundations of thermodynamics. It uses *phenomena* displayed by photons, atoms, etc. to provide the data to which his argument is applied, but does not rely on a quantum mechnical description of those phenomena. In particular, there are circumstances, first identified by Einstein, Podolsky and Rosen (EPR) in 1935, where deep intuitions about locality and cause and effect seem to demand, whether the underlying theoretical description is classical, quantum mechanical, or still undiscovered, that certain kinds of measurements reveal pre-existing values. Bell's Theorem applies directly to such experiments, and says that if their results are as predicted by quantum mechanics, then those pre-existing values ('hidden variables' only from the point of view of quantum mechanics) cannot exist. Bell's thoughts on the disquieting nature of this result are noteworthy [6]:

> For me, it is so reasonable to assume that the photons in those [EPR] experiments carry with them programs, which have been correlated in advance, telling them how to behave. This is so rational that I think that when Einstein saw that, and the others refused to see it, *he* was the rational man. The other people, although history has justified them, were burying their heads in the sand . . . Einstein's intellectual superiority over Bohr, in this instance, was enormous; a vast gulf between the man who saw clearly what was needed, and the obscurantist. So for me, it is a pity that Einstein's idea doesn't work. The reasonable thing just doesn't work.

This then brings us to Bell's more general views regarding quantum mechanics. His skeptical attitude towards quantum mechanics did not stem from the counter-intuitive implications of the theory's successful passage of experimental tests that evolved from his Theorem. He had expected that outcome, and – unlike Epstein – did not find utterly unacceptable the kind of non-locality it implied. Bell doubted that the theory was simply wrong, 'but he *knew* it was rotten', and in saying so he liked to pronounce 'rotten' with gusto [7]. What he found rotten was the formulation of the theory [8]:

> . . . the quantum mechanical description will be superseded. In this it is like all theories made by man. But to an unusual extent its ultimate fate is apparent in its internal structure. It carries in itself the seeds of its own destruction.

He found especially offensive the reliance of that formulation on the notion of 'measurement'. Indeed, his last article is called *Against 'measurement'* [9], and advocates that 'the word [measurement] has had such a damaging effect [that] it should be banned[3] altogether in quantum mechanics'. To characterize standard treatments of 'measurement' he introduced the biting acronym FAPP – 'for all practical purposes', which he occasionally amplified to QED FAPP to point to proofs he found to rest on 'fuzzy logic'.

Discussion of 'measurement' has no place in the *formulation* of *any* fundamental theory, he argued, for this put the cart before the horse by introducing 'an artificial division of the world, and an intention to neglect, or to take only a schematic [i.e., FAPP] account of, the interaction across the split'. He admonished his colleagues to stay true to what he saw as the physicist's

[3] Other words that Bell sought to ban from the *formulation* of the theory (as compared to discussions of its applications) are *system, apparatus, environment, microscopic, macroscopic, reversible, irreversible, information* and *observables*, the latter term to be replaced by one of his favorite concepts, *beables*.

duty [10]:

> In the beginning natural philosphers tried to understand the world around them . . . they hit upon the idea of converting artificially simple situations . . . Divide and conquer. Experimental science was born. But experiment is a tool. The aim remains: to understand the world. To restrict quantum mechanics to be exclusively about piddling laboratory operations is to betray the great enterprise.

And he proclaimed that those who seek to advance this enterprise must obey Einstein's dictum 'that it is the theory which decides what is observable', and not the other way around.

Having concluded that no interpretation of orthodox quantum mechanics could be devised that he would find acceptable, Bell was more than willing to countenance radical revisions. In responding to whether he [11] 'would prefer to retain the notion of objective reality and throw away one of the tenets of relativity', he answered

> Yes. One wants to be able to take a realistic view of the world as if it is really there, even when it is not being observed . . . [I] want to go back to the idea of an aether . . . , because in these EPR experiments there is a suggestion that behind the scenes something is going faster than the speed of light . . . And so it's precisely to avoid these [paradoxes of causality] that I want to say that there is a real causal sequence which is defined in the aether . . . It is as if there is some kind of conspiracy, that something is going on behind the scenes which is not allowed to appear.

His assault on the 'infamous "measurement" problem' led him to be intrigued by proposals [12] which postulate nonlinear stochastic modifications of Schrödinger's equation specifically designed to collapse the wave function to *one or another* of the outcomes of the orthodox theory in a time that, for simple systems, is sufficiently long to leave the standard predictions intact, but which assures us that [13] 'pointers very rapidly point, and cats are very quickly killed *or* spared'.

Bell has had the greatest impact on the interpretation of quantum mechanics of anyone since the 1920s. He belonged, also, to that small company of physicists whom either of us would walk miles to hear lecture on any topic whatever. Bell spoke softly, but with intensity and passion, and explained matters of great subtlety with consummate skill. His wit was sparkling, but he also displayed something like the wrath of the Old Testament prophet for those who adhered to positions he judged superficial. He responded to challenging questions in beautifully formed, concise and simple sentences. The unforgettable music of his Irish voice was surely a part of the magic, but we can demonstrate that there was far more to it by letting him speak for himself [14]:

> In my opinion, these views are too complacent.[4] The pragmatic approach which they exemplify has undoubtedly played an indispensable role in the evolution of contemporary physical theory. However, the notion of the 'real' truth, as distinct from a truth that is presently good enough for us, has also played a positive role in the history of science. Thus Copernicus found a more intelligible pattern by placing the sun rather than the earth at the center of the solar system. I can well imagine a future phase in which this happens again, in which the world becomes intelligible to human beings, even to theoretical physicists, when they do not imagine themselves to be the center of it.

It was our good fortune to have spent a week in June 1990 with John and Mary in a workshop at Amherst College, where these issues were discussed at leisure and at length. Afterwards, driving back to Ithaca, we agreed that John was truly unique in the world of physics, as a personality and as an intellect – at once scientist, philosopher and humanist. He was a person to whom deep ideas mattered deeply. Fate has been most cruel to steal him from us when he was still so brimful of vitality. But he will live on through his profound and timeless work. That, and the privilege of having known him, must be our solace.

REFERENCES

[1] J. Bernstein, *Quantum Profiles* (Princeton University Press, 1991).

[2] J.S. Bell, *Speakable and Unspeakable in Quantum Mechanics* (Cambridge University Press, 1987).

[3] J.S. Bell, On the problem of hidden variables in quantum mechanics, *Rev. Mod. Phys.* **38** (1966) 447.

[4] J.S. Bell, On the Einstein–Rosen–Podolsky paradox, *Physics* **1** (1964) 195. This and the preceding paper were actually written at about the same time.

[5] Ref. [2], p. 20.

[4] Bell and Nauenberg expressed this same point with zest in Ref. [8]: 'We emphasize not only that our view is that of a minority, but also that current interest in such questions is small. The typical physicist feels that they have long been answered, and that he will fully understand just how if ever he can spare just twenty minutes to think about it.'

[6] Bernstein, ref. [1], p. 84.
[7] Bernstein, ref. [1], p. 20.
[8] From an essay by J.S. Bell and M. Nauenberg, 'The moral aspect of quantum mechanics', ref. [2], p. 27.
[9] J.S. Bell, Against "measurement", *Physics World*, August 1990, pp. 33–40.
[10] Ref. [9], p. 34.
[11] J.S. Bell, in *The Ghost of the Atom*, P.C.W. Davies and J.R. Brown (eds.) (Cambridge University Press, 1986), p. 50.
[12] See, in particular, G.C. Ghirardi, A. Rimini and T. Weber, *Phys. Rev.* **D34** (1986) 470; **D36** (1987) 3287 and ref. [2], pp. 201–212.
[13] Ref. [9], p. 40. Note the stress here on *or* in contrast to *and.*
[14] Ref. [2], p. 125.

9 Remembering John Bell*

ROMAN JACKIW

John Bell and I met and became acquainted in 1967, when I went to CERN for a year-long research visit, soon after finishing my doctoral studies at Cornell. At that time, particle physics theory was dominated, as it happens from time-to-time, by a single idea; there was broad agreement among theorists what the important problems are and how they should be solved – these days one hardly remembers the details of that program. But attaching my scientific activity to a consensus was not my ambition; I had much admired the independent attitude of one of my research supervisors at Cornell, Ken Wilson. So I looked among the staff at CERN for someone who pursued interesting issues that were neither 'central' nor 'important', and I was delighted to find such a scientist in John Bell. Moreover, he was generous in giving his time; he tolerated my coming to his office and appeared willing to discuss without limit. I appreciated the magnitude of his generosity only years later when I too became installed in an office and people began coming in and taking my time to talk about things.

There began for us a period of wide-ranging conversations, not only about physics, which acquainted me with the many issues that concerned John; though nothing was then said about his work on quantum mechanics – he did not at that time describe it to me and I did not know of it. Current algebra interested John very much. Within its framework one can understand the low-energy behavior of elementary particles, without making a commitment to a specific dynamical model, which in the 1960s was unknown, while today's 'standard model' resists solution in the low-energy domain. The approach seemed successful, complete and exhausted by the late 1960s, yet there remained discrepancies between theoretical prediction and experimental verification.

* This work is supported in part by funds provided by the U.S. Department of Energy (D.O.E.) under contract #DE-AC02-76ER03069.

John was particularly impressed with an analysis by his good friend M. Veltman, and also D. Sutherland, to the end that the neutral pion could not decay into two photons, if the charge-neutral and gauge invariant axial vector (chiral) current is conserved, as it is taken to be in current algebra applications. Because in fact the decay does occur in Nature, while the Sutherland/Veltman argument appeared incontrovertible, John stressed that the subject of current algebra must not be closed until this puzzle is resolved, and urged studying the chiral current.

This was the second time I received such advice: in my final student days Wilson suggested a critical examination of the apparent conservation of the axial vector current in the Barker–Johnson–Willey theory of massless electrodynamics, with which he had his own disagreements.

Thus I was very willing to research this topic, but since the existing discussions were straightforward and the conclusions immediate, it was hard to see how a useful probe could be launched. So I asked my fellow theorists for suggestions but the subject did not spark interest. I do recall two mathematically oriented colleagues, Henri Epstein and Raymond Stora, offering a diagnosis that in retrospect proved prescient: in their opinion one could not rely on current algebra analyses because physicists treat cavalierly singular products of distributions. But their prognosis that a cure will be found if one uses rigorous rather than heuristic mathematics did not appeal to me. In fact, the decisive suggestion did not come from a theorist but from an experimentalist.

One of the civilized activities at CERN, to which John frequently invited me, consisted of taking an afternoon drink at the cafe, where we would continue our conversations together with people who joined us. One time, Jack Steinberger – John's friend and collaborator on a CP formalism – was at the table and asked about our current interests. When he described to him the $\pi^0 \to 2\gamma$ puzzle, he expressed amazement that theorists should still be pursuing a process that he, an experimentalist, calculated almost twenty years earlier, finding excellent agreement with experiment, while also noting a discrepancy between results obtained when the pion coupled to nucleons by pseudovector or pseudoscalar interactions. (Pions, nucleons and photons were the only particles in Steinberger's model, and it was believed that equivalent results emerge for pseudovector and pseudoscalar pion–nucleon coupling.)

There at that table came to us the realization that Steinberger's calculation would be identical to the one performed in the dynamical framework of the

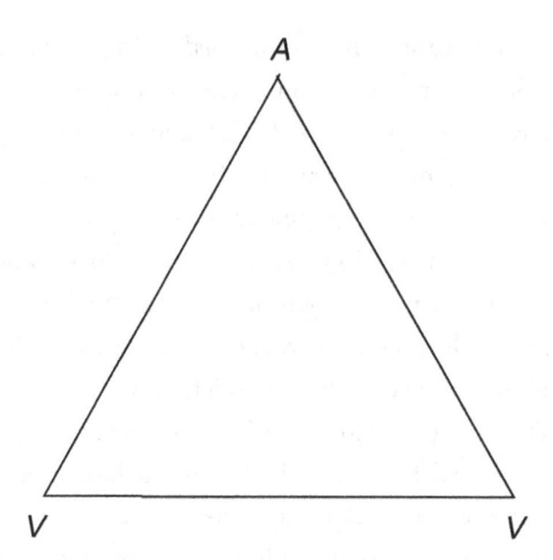

FIGURE 9.1 Triangle Feynman diagram that leads to quantum mechanical symmetry breaking. Two vector currents V and one axial current A form the three corners, at which virtual fermions are created/annihilated. Propagation of the virtual fermions, indicated by solid lines, prevents the currents from being conserved. A similar effect arises with three axial vector currents.

σ-model, which was constructed to realize current algebra explicitly. We reasoned that within the σ-model we could satisfy the current algebraic assumptions of Sutherland/Veltman and also obtain good experimental agreement in view of Steinberger's result – thereby resolving the $\pi^0 \rightarrow 2\gamma$ puzzle.

Guided by Steinberger's paper (at that time we were not familiar with the work of Steinberger's contemporaries, H. Fukuda and Y. Miyamoto, and only dimly aware of subsequent contributions by J. Schwinger), we quickly established that the correlation amplitude for the three currents of the problem, two vector currents to which the two photons couple and one axial vector current to which the pion couples, is given in lowest order (one loop) perturbation theory by the now famous triangle graph depicted in Fig. 9.1, whose value is determined by Feynman rules only up to an overall ambiguity, owing to ultraviolet divergences, even though the amplitude is finite. Moreover, while the ambiguity may be resolved by enforcing current conservation, it is impossible to maintain conservation of all *three* currents, as is assumed in the current algebra calculation. Thus we found that the σ-model's symmetries, which underlie current algebra and which should guarantee the conservation of the respective currents, cannot be maintained when the model is quantized. In the absence of these symmetries, pion decay is no longer forbidden.

Our work resolved the $\pi^0 \to 2\gamma$ problem, by exposing a purely quantum mechanical mechanism for symmetry breaking, which is the modern interpretation of Steinberger's discrepancy, and these days is described as 'anomalous breaking of a symmetry', although once the surprise has worn off, it is better named 'quantum mechanical symmetry breaking' [1].

Our analysis of the 'anomaly' was complemented by S. Adler [2], who, working independently, came to a similar conclusion about (absence of) symmetries in massless electrodynamics, and, building on our work, established, with W. Bardeen, the important fact that higher perturbative orders do not modify the one-loop calculation of pion decay. Further confirmation came from Wilson, who used our theory as a case study for his non-Lagrangian models of current algebra, based on his operator product expansion. The early period of research on this subject culminated in a phenomenological description of quantum mechanical symmetry breaking in terms of an effective Lagrangian, constructed by J. Wess and B. Zumino (who apparently were not aware of our result) [3].

In time the work, which arose from clearing up a corner of current algebra, grew to affect much of particle physics. It became an important ingredient of model building, both for speculative strings and for the conventional 'standard model', where, among other things, it enforces color triality and explains the numerical equality of quark and lepton degrees of freedom, thus predicting the existence of the top quark [3].

John maintained an amused interest as our calculation became transformed in various contexts, and was shown to be a consequence of diverse physical and mathematical considerations [3]: symmetry breaking aspects of the Dirac sea (Feynman), anomalous transformation properties of the functional integral measure (Fujikawa), the necessary effect of high-energy modes on low-energy physics (Gribov), quantum field theoretic manifestation of Berry's phase, local version of the Atiyah–Singer index, and cohomological properties of gauge groups (Faddeev). The last two mathematical connections seeded a remarkable collaboration between mathematics and physics, which is still flourishing. On the other hand, the physical world itself became threatened by the anomaly because, as G. 't Hooft showed [4], it catalyzes baryon decay, but fortunately at a sufficiently slow rate to cause no immediate concern.

In spite of these wide-ranging generalizations, John preferred the simple triangle graph calculation [1]. He always stressed the element of choice that exists in resolving the calculational ambiguity, thus putting different faces on the nature of the anomaly – a freedom that is obscured in the more abstract and high-powered approach.

Indeed John was rather diffident about the entire matter. In this he showed one of his many striking qualities: modesty about his own work, praise for the work of others, but skepticism in the face of inflated claims, even if they were extolling his own contributions. When he eventually described to me his famous analysis of quantum measurement theory [5], he called that research 'a hobby'.

After I left CERN in 1968, we had many occasions to meet and talk about interesting topics, but our discussions never again resulted in a joint publication. The closest we came to this happened when the phenomenon of fractional quantum numbers [6] became physically relevant [7]. John found this interesting but characteristically was at first skeptical that a fractional value could be an eigenvalue. Upon elaborating the precise circumstances in which a sharp observable arises, he published with R. Rajaraman [8] an analysis that contributed to the understanding and acceptance of this fascinating idea, which today has also gained wide currency.

In all my contacts with John I was always made aware of his overwhelmingly intellectual precision and honesty. These are the qualities that made him such an incisive critic and therefore a wonderful colleague. Moreover, this attitude lay behind his scientific achievements, which are informed by clarity of observation about previously murky subjects.

The same attitude characterized his approaches outside science, for example to social and political questions. Many physicists profess humane and liberal values, but often these become obscured by personal emotion and prejudice. In the last quarter century, issues of Vietnam, Ireland and Palestine offer a dramatic opportunity for displaying social conscience in search of justice. John recognized and spoke on these matters clearly. Still in the late 1960s I heard him analyze America's role in Vietnam in terms that did not gain acceptance until years later; his opinions on the two other tragedies remain in the minority even today, but one hopes that here too his ideas are merely ahead of their time.

I liked John very much and together with many colleagues I shall miss him. He was an outstanding scientist and helped us do good science, which is one reason why we become physicists. Moreover, many enter our field not only for the opportunity of exploring Nature in its most fundamental workings, but also for what we perceive as the purity and honesty of the profession. These qualities sometimes get submerged by pressure of personal ambition, struggling for achievement and recognition, but John Bell never lost them, and in this way he reminded us of the other reason for becoming a physicist.

REFERENCES

[1] J.S. Bell and R. Jackiw, A PCAC puzzle: $\pi^0 \rightarrow 2\gamma$ in the σ model, *Nuovo Cimento* **60** (1969) 47.

[2] S.L. Adler, Axial-vector vertex in spinor electrodynamics, *Phys. Rev.* **177** (1969) 2426.

[3] For detailed discussion and guide to the literature see S. Trieman, R. Jackiw, B. Zumino and E. Witten, *Current Algebra and Anomalies* (Princeton University Press/World Scientific, Princeton, NJ/Singapore, 1985).

[4] G. 't Hooft, Symmetry breaking through Bell–Jackiw anomalies, *Phys. Rev. Lett.* **37** (1976) 8.

[5] J.S. Bell, *Speakable and Unspeakable in Quantum Mechanics* (Cambridge University Press, Cambridge, UK, 1987).

[6] R. Jackiw and C. Rebbi, Solitons with fermion number 1/2, *Phys. Rev.* **D13** (1976) 3398.

[7] W.-P. Su, J.R. Schrieffer and A. Heeger, Solitons in Polyacetylene, *Phys. Rev. Lett.* **42** (1979) 1968; R. Jackiw and J. R. Schrieffer, Solitons with fermion number 1/2 in condensed matter and relativistic field theories, *Nucl. Phys.* **B190** [FS3] (1981) 253.

[8] R. Rajaraman and J.S. Bell, On solitons with half-integer charge, *Phys. Lett.* **116B** (1982) 115; J. S Bell and R. Rajaraman, On states, on a lattice, with half-integer charge, *Nucl. Phys.* **B220** [FS8] (1983) 1.

Index

Printed in the United States
By Bookmasters